METHODS IN MOLECULAR BIOLOGY™

Series Editor
John M. Walker
School of Life Sciences
University of Hertfordshire
Hatfield, Hertfordshire, AL10 9AB, UK

METHODS IN MOLECULAR BIOLOGY

Series Editor
John M. Walker
School of Life Sciences
University of Hertfordshire
Hatfield, Hertfordshire, AL10 9AB, UK

For further volumes published in this series, go to
www.springer.com/series/7651

Lentivirus Gene Engineering Protocols

Second Edition

Edited by

Maurizio Federico

Division of Pathogenesis of Retroviruses, National AIDS Center,
Istituto Superiore di Sanità, Rome, Italy

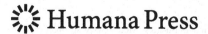

Editor
Maurizio Federico, Ph.D.
Division of Pathogenesis of Retroviruses
National AIDS Center
Istituto Superiore di Sanità
Rome
Italy
maurizio.federico@iss.it

ISSN 1064-3745   e-ISSN 1940-6029
ISBN 978-1-4939-5653-1   ISBN 978-1-60761-533-0 (eBook)
DOI 10.1007/978-1-60761-533-0
Springer New York Dordrecht Heidelberg London

Cover illustration: Derived from Chapter 9, figure 1A.

Printed on acid-free paper

Humana Press is part of Springer Science+Business Media (www.springer.com)

Dedicated to my wife, Margherita, and to the memory of my mother, Ottavia. As a token of my gratitude for their love.

Preface

This second edition of the *Lentivirus Gene Engineering Protocols* book does not simply update several aspects of the techniques detailed in the previous edition published in 2003. Rather, it mirrors the spectacular applicative progress which has occurred in the field of lentiviral vector (LV)-based gene engineering over the past few years. On the other hand, considering that many LV-related techniques (e.g., LV titration and challenge, detection and selection of transduced cells, detection of replication-competent viral particles) did not meet relevant modifications over the years, readers interested in these and other matters not included in the present book can still conveniently refer to the previous edition. Conversely, an almost new set of topics is included in this second edition. The only exception is the contribution from Kamen et al. who describe a significantly improved method for LV production relying on the use of a highly transfectable HEK293 cell line growing in suspension and in serum-free conditions. This reagent offers the possibility of setting high density LV-producing cell cultures, as well as recovering high amounts of LV particles free of fetal serum contaminants.

This volume is introduced by a very comprehensive review in which Dreyer covers basically all the extraordinary advances allowed by the use of LVs in the field of biology and dysfunctions of the nervous system. In Part II of the book, new responses toward a number of key issues concerning LV use are provided. In particular, Bischof and Cornetta illustrate concepts and techniques for the best choice of receptors to be inserted in LV particles, considering that some receptors have the potential to infect a wide variety of cell types, while other viral envelope glycoproteins can facilitate specific cell targeting. On the other hand, for many experimental and clinical applications, regulated expression of the transgene is often desirable. In this regard, in a dedicated chapter, Markusic and Seppen offer a guide to select the best doxycycline inducible LV for different applications. However, in some cases, the transgene can undergo silencing at variable times after integration in host DNA. In this regard, although LVs significantly ameliorate the outcomes obtained using gammaretroviral vectors, this still remains a significant shortcoming mainly for applications requiring long lasting transgene expression (e.g., treatment of degenerative diseases). Ramezani and Hawley describe the use of various genetic regulatory elements which, once inserted into LVs, minimize transgene silencing and other effects induced by neighboring chromatin. An alternative strategy considers the insertion of a transgene of interest into non-integrating LVs, as detailed by Cara et al. Part II is completed by the description of an original method of protein delivery based on the high incorporation levels in HIV-1 Virus-like particles of an HIV-1 Nef mutant which can act as an efficient anchoring element for heterologous proteins.

Part III is devoted to the description of the newest and most relevant applications of LV-based gene engineering. In particular, Du and Zhang describe a LV-mediated method to establish stable transgenic human embryonic stem cells, while Chong and Chan report the techniques that would be on the basis of upcoming LV-based fetal gene therapy for the treatment of inherited deficiencies. Further, Bonci et al. describe how LV-base gene engineering can efficiently manipulate the cell differentiation program. Finally, Chang demonstrates the feasibility of efficient vaccine designs based on LV-mediated transduction of dendritic cells.

Part IV of the book is entirely dedicated to a new and invaluable application field, i.e., the LV-mediated expression of micro (mi)RNAs. These are highly conserved small regulatory RNAs, which regulate gene expression by hybridization to specific binding sites in the 3′ untranslated region (UTR) of many mRNAs. On this subject, Scherr and colleagues describe the expression of individual miRNAs or miRNA-specific antagomiRs by LV gene transfer to induce stable gain- and loss-of function phenotypes. Specific applications of LV-expressed miRNAs in the fields of oncology (by Sumimoto and Kawakami), HIV-1 infection (by Berkout et al.), and neurodegenerative diseases (by Peng and Masliah) are also comprehensively treated in the following chapters.

Overall, this book reports almost all new LV-based gene engineering techniques, which are described by outstanding scientists actively involved in the field. This and the previous edition of *Lentivirus Gene Engineering Protocols* cover the most relevant issues and techniques of LV-based gene engineering, thus representing a complete theoretical/practical guide for scientists still unfamiliar with LV technologies.

As with the first edition, I'm pleased to acknowledge the secretarial assistance of F.M. Regini.

Maurizio Federico

Contents

Contributors

SVEN ANSORGE • *Biotechnology Research Institute, NRC, Montreal, QC, Canada*

BEN BERKHOUT • *Laboratory of Experimental Virology, Department of Medical Microbiology, Center for Infection and Immunity Amsterdam (CINIMA), University of Amsterdam, Amsterdam, The Netherlands*

DANIELA BISCHOF • *Department of Medical and Molecular Genetics, Indiana University School of Medicine, Indianapolis, IN, USA*

ROBERTA BONA • *Division of Pathogenesis of Retroviruses, National AIDS Center, Istituto Superiore di Sanità, Rome, Italy*

DÉSIRÉE BONCI • *Department of Hematology, Oncology, and Molecular Medicine, Istituto Superiore Sanità, Rome, Italy*
Reproductive Technologies Laboratory, Avantea, Cremona, Italy

ANDREA CARA • *Department of Drug Research and Evaluation, Istituto Superiore di Sanità, Rome, Italy*

JERRY CHAN • *Experimental Fetal Medicine Group, Department of Obstetrics and Gynecology, Yong Loo Lin School of Medicine, National University of Singapore, Singapore*

LUNG-JI CHANG • *Department of Molecular Genetics and Microbiology, Powell Gene Therapy Center and McKnight Brain Institute, College of Medicine, University of Florida, Gainesville, FL, USA*

MARK S.K. CHONG • *Experimental Fetal Medicine Group, Department of Obstetrics and Gynecology, Yong Loo Lin School of Medicine, National University of Singapore, Singapore*

VALERIA COPPOLA • *Department of Hematology, Oncology, and Molecular Medicine, Istituto Superiore Sanità, Rome, Italy*
Reproductive Technologies Laboratory, Avantea, Cremona, Italy

KENNETH CORNETTA • *Department of Medical and Molecular Genetics, Indiana University School of Medicine, Indianapolis, IN, USA*

JEAN-LUC DREYER • *Division of Biochemistry, Department of Medicine, University of Fribourg, Fribourg, Switzerland*

ZHONG-WEI DU • *Departments of Anatomy and Neurology, School of Medicine and Public Health, WiCell Institute, University of Wisconsin, Madison, WI, USA*

YVES DUROCHER • *Biotechnology Research Institute, NRC, Montreal, QC, Canada*

MATTHIAS EDER • *Department of Hematology, Hemostasis, Oncology, and Stem Cell Transplantation, Hannover Medical School, Hannover, Germany*

MAURIZIO FEDERICO • *Division of Pathogenesis of Retroviruses, National AIDS Center, Istituto Superiore di Sanità, Rome, Italy*

CESARE GALLI • *Department of Hematology, Oncology, and Molecular Medicine, Istituto Superiore Sanità, Rome, Italy*
Reproductive Technologies Laboratory, Avantea, Cremona, Italy

ALAIN GARNIER • *Department of Chemical Engineering, Centre de Recherche sur la Fonction, la Structure et l'Ingénierie des Protéines, Université Laval, Québec City, QC, Canada*

ROBERT G. HAWLEY • *Department of Anatomy and Regenerative Biology, The George Washington University Medical Center, Washington, DC, USA*

AMINE KAMEN • *Biotechnology Research Institute, NRC, Montreal, QC, Canada*

YUTAKA KAWAKAMI • *Division of Cellular Signaling, Institute for Advanced Medical Research, Keio University School of Medicine, Tokyo, Japan*

DAVID MARKUSIC • *AMC Liver Center, Amsterdam, The Netherlands*

ELIEZER MASLIAH • *Department of Neurosciences and Department of Pathology, University of California, San Diego, La Jolla, CA, USA*

MARÍA MERCEDES SEGURA • *Biotechnology Research Institute, NRC, Montreal, QC, Canada; Department of Chemical Engineering, Centre de Recherche sur la Fonction, la Structure et l'Ingénierie des Protéines, Université Laval, Québec City, QC, Canada; Center of Animal Biotechnology and Gene Therapy (CBATEG) Universitat Autònoma de Barcelona, Barcelona, Spain*

ZULEIKA MICHELINI • *Department of Drug Research and Evaluation, Istituto Superiore di Sanità, Rome, Italy*

CLAUDIA MURATORI • *Division of Pathogenesis of Retroviruses, National AIDS Center, Istituto Superiore di Sanità, Rome, Italy*

MARIA MUSUMECI • *Department of Hematology, Oncology, and Molecular Medicine, Istituto Superiore Sanità, Rome, Italy Reproductive Technologies Laboratory, Avantea, Cremona, Italy*

DONATELLA NEGRI • *Department of Infectious, Parasitic, and Immune-Mediated Diseases, Istituto Superiore di Sanità, Rome, Italy*

KEVIN A. PENG • *School of Medicine, University of California, San Diego, La Jolla, CA, USA*

ALI RAMEZANI • *Department of Anatomy and Regenerative Biology, The George Washington University Medical Center, Washington, DC, USA*

MICHAELA SCHERR • *Department of Hematology, Hemostasis, Oncology, and Stem Cell Transplantation, Hannover Medical School, Hannover, Germany*

JURGEN SEPPEN • *AMC Liver Center, Amsterdam, The Netherlands*

HIDETOSHI SUMIMOTO • *Division of Cellular Signaling, Institute for Advanced Medical Research, Keio University School of Medicine, Tokyo, Japan*

OLIVIER TER BRAKE • *Laboratory of Experimental Virology, Department of Medical Microbiology, Center for Infection and Immunity Amsterdam (CINIMA), University of Amsterdam, Amsterdam, The Netherlands*

LETIZIA VENTURINI • *Department of Hematology, Hemostasis, Oncology, and Stem Cell Transplantation, Hannover Medical School, Hannover, Germany*

JAN-TINUS WESTERINK • *Laboratory of Experimental Virology, Department of Medical Microbiology, Center for Infection and Immunity Amsterdam (CINIMA), University of Amsterdam, Amsterdam, The Netherlands*

SU-CHUN ZHANG • *Departments of Anatomy and Neurology, School of Medicine and Public Health, WiCell Institute, University of Wisconsin, Madison, WI, USA*

Part I

Overview

Chapter 1

Lentiviral Vector-Mediated Gene Transfer and RNA Silencing Technology in Neuronal Dysfunctions

Jean-Luc Dreyer

Abstract

Lentiviral-mediated gene transfer in vivo or in cultured mammalian neurons can be used to address a wide variety of biological questions, to design animal models for specific neurodegenerative pathologies, or to test potential therapeutic approaches in a variety of brain disorders. Lentiviruses can infect nondividing cells, thereby allowing stable gene transfer in postmitotic cells such as mature neurons. An important contribution has been the use of inducible vectors: the same animal can thus be used repeatedly in the doxycycline-on or -off state, providing a powerful mean for assessing the function of a gene candidate in a disorder within a specific neuronal circuit. Furthermore, lentivirus vectors provide a unique tool to integrate siRNA expression constructs with the aim to locally knockdown expression of a specific gene, enabling to assess the function of a gene in a very specific neuronal pathway. Lentiviral vector-mediated delivery of short hairpin RNA results in persistent knockdown of gene expression in the brain. Therefore, the use of lentiviruses for stable expression of siRNA in brain is a powerful aid to probe gene functions in vivo and for gene therapy of diseases of the central nervous system. In this chapter, I review the applications of lentivirus-mediated gene transfer in the investigation of specific gene candidates involved in major brain disorders and neurodegenerative processes. Major applications have been in polyglutamine disorders, such as synucleinopathies and Parkinson's disease, or in investigating gene function in Huntington's disease, dystonia, or muscular dystrophy. Recently, lentivirus gene transfer has been an invaluable tool for evaluation of gene function in behavioral disorders such as drug addiction and attention-deficit hyperactivity disorder or in learning and cognition.

Key words: miRNA, Lentivirus, Brain diseases, Gene therapy, mRNA silencing

1. Introduction

Lentiviral vectors are increasingly used for gene delivery to neurons and in experimental models of neurodegeneration. Tackling neurodegenerative diseases represents a formidable challenge for our aging society (1, 2). Local delivery of therapeutic molecules represents one of the limiting factors for the treatment of

Maurizio Federico (ed.), *Lentivirus Gene Engineering Protocols,* Methods in Molecular Biology, vol. 614,
DOI 10.1007/978-1-60761-533-0_1, © Humana Press, a part of Springer Science + Business Media, LLC 2010

neurodegenerative disorders. Gene transfer methods for efficient coexpression of exogenous proteins in neurons are crucial tools toward the understanding of the molecular basis of the central nervous system. In vivo gene transfer using lentiviral vectors constitutes a powerful tool in this area (3). Major advantages of lentiviral vectors are their ability to transduce nondividing cells, including differentiated neurons, and to confer long-term expression of transgenes (4, 5).

In recent years, the breakthrough discovery that double-stranded RNA of 21 nucleotides in length (referred to as short or small interfering RNA; siRNA) triggers sequence-specific gene silencing in mammalian cells led to the development of a powerful new approach to study gene function (6–8). This evolutionarily conserved mechanism is carried out by an endogenous pathway that centers on the use of endogenously encoded small RNAs, and can be hijacked to knockdown the expression of any target protein by introducing a specific siRNA into a cell (9). Therefore, the ability to manipulate the RNA-induced interference (RNAi) in mammalian cells provides a very powerful tool to influence gene expression. Stable knockdown can be obtained by constitutive expression of the siRNA from the host chromosome, but neurons display resistance to transduction by siRNAs, when attempting to silence expression of endogenous genes in these postmitotic cells. Thus, for in vivo applications, RNAi has been hampered until recently by inefficient delivery methods and by the transient nature of the gene suppression. However, lentivirus-mediated expression of short hairpin (sh)RNA was shown to be long term and mediated stable RNAi in a dose-dependent manner (10), therefore providing a convenient way to integrate RNAi expression constructs. Besides their ability to infect nondividing cells, thereby allowing knockdown in mature neurons, lentiviruses also improve the efficiency of gene delivery to different cell types, thus improving the potential usefulness of RNAi as therapy (10–14). A number of lentiviral vectors have been developed for performing sh-directed RNAi in mice (13, 15–18). Some vectors encode green fluorescent protein (GFP) as a marker for infectivity and to track the RNAi knockdown cells in real-time. Other vectors contain antibiotic selection markers, such as puromycin, which are useful when the cell selection is imposed. In addition, there are conditional versions of lentiviral RNAi vectors available, where the user may opt to 'turn on' or 'turn off' RNAi during the course of experimentation (4, 15, 18). Drug-inducible systems allowing the control of transgene expression and knockdown in mammalian cells are invaluable tools for genetic research, and could also play important roles in translational research or gene therapy (19). A lentivector-based, conditional gene expression system for drug-controllable expression of transgenes and shRNAs has been established. This is a highly robust and versatile system governing tightly controlled expression of both transgenes and endogenous cellular genes in various primary and established cell lines in vitro, as well as in vivo in the central nervous system (19). The Woodchuck hepatitis virus (WHV) posttranscriptional regulatory element (WPRE) facilitates nucleocytoplasmic transport of RNA mediated by several alternative pathways that may be cooperative and this element has been included in many different gene therapy vectors (e.g., lentiviral vectors, AAV, adenovirus) to stimulate heterologous cDNAs expression. More recently, insertion of two copies of a human synapsin promoter/WPRE cassette in a single lentiviral vector was shown to direct robust co-expression of cDNAs in cultured neurons, while excluding expression in the surrounding glial cells (5). In addition, insertion of the tetracycline-inducible system (Tet-off) controlled by the synapsin promoter results in tightly regulated expression of EGFP when used as a transgene in cultured neurons. This provides a neuron-specific and regulated expression mediated by single lentiviral vectors and represents a valuable tool for the study of neuronal functions (5).

A major breakthrough in our understanding of specific gene functions in many brain disorders has been made possible by either overexpressing or silencing gene expression in very specific neuronal pathways. For this purpose, the use of replication uncompetent viruses preventing the vector to spread away from the injection area, has been a major advance. Lentiviruses have the additional advantage that large gene can be inserted and permanently incorporated into the host cell. Furthermore, the use of neuron-specific promoters enables cell-specific gene transfer in a complex system such as the brain. In this review, I will discuss the applications of lentivirus-mediated gene transfer in the investigation of a large variety of brain disorders, with particular emphasis on the use of lentiviruses for local gene knockdown.

2. Multiple Sclerosis

Multiple sclerosis (MS) is an autoimmune condition in which the immune system attacks the central nervous system, leading to demyelination (20, 21). Disease onset usually occurs in young adults (22) with a prevalence that ranges between 2 and 150 per 100,000 (23). Multiple sclerosis affects the white matter of the brain and spinal cord, destroying the myelin sheath and producing lesions (scleroses or plaques) in the white matter (22). Although much is known about the mechanisms involved in the disease process, the cause remains unknown. Theories include genetics or infections (24). It is not considered a hereditary disease, however, specific genes have been linked with multiple sclerosis (25) and genetics may play a role in determining a person's susceptibility to multiple sclerosis.

Differences in the human major histocompatibility complex increase the probability of suffering multiple sclerosis (26).

Besides, mutations in the IL-2RA and the IL-7RA subunits of the receptor for interleukin 2 and interleukin 7, respectively, are associated with the disorder (27, 28). The histocompatibility complex is involved in antigen presentation, while mutations in the IL-2 and IL-7 genes were already known to be associated with diabetes and other autoimmune conditions, supporting the notion that multiple sclerosis is an autoimmune disease (26, 29, 30). Other studies have linked genes in chromosome 5 with the disease (31). Richardson et al. (32) have used intrabody-mediated knockout of the high-affinity IL-2 receptor using a bicistronic lentivirus vector, but this has not yet been evaluated for the disease.

Furthermore, several studies demonstrate that lesions from patients contain activated transcription factors like NF-κB, STAT1, STAT3, and STAT6 (33–35), which can lead to enhanced signaling of inflammatory stimuli. Thus, regulation of inflammatory signaling may be altered in MS, and may be responsible for inflammatory demyelination. SHP-1 is a protein tyrosine phosphatase with two SH2 domains that acts as a crucial negative regulator of cytokine signaling, inflammatory gene expression, and demyelination in central nervous system via STAT1, STAT3, and STAT6 (36–39). Mice genetically lacking in SHP-1 (motheaten mice) display myelin deficiency, which may be mediated by increased inflammatory mediators in the CNS (40, 41). Christophil et al. (42) established a role for SHP-1 in MS. The levels of SHP-1 protein and mRNA in peripheral blood mononuclear cells (PBMCs) of patients were significantly lower compared to normal subjects. Moreover, promoter II transcripts, expressed from one of two known promoters, were selectively deficient in MS patients. As expected, patients had significantly higher levels of pSTAT6. Furthermore, siRNA to SHP-1 effectively increased the levels of phosphorylated pSTAT6 in PBMCs of controls to levels equal to patients. Christophil et al. (42) also designed a lentiviral vector expressing SHP-1 (43). The vector carried the human SHP-1 coding sequence, allowing the bicistronic expression of green fluorescent protein (GFP) and SHP-1 in the transduced cells. Transduction of PBMCs with this lentivirus lowered pSTAT6 levels. Multiple STAT6-responsive inflammatory genes were increased in PBMCs of patients relative to PBMCs of normal subjects. Thus, lentiviruses may be very efficient tools to assess the mechanisms of inflammatory demyelination.

3. Dystonia

Dystonia is a neurological movement disorder in which sustained muscle contractions cause twisting and repetitive movements or abnormal postures (44). The disorder may be inherited or caused by other factors such as birth-related or other physical trauma, infection, poisoning (e.g., lead poisoning) or reaction to drugs. Primary dystonia is suspected to be caused by a pathology of the central nervous system, likely originating in those parts of the brain concerned with motor function, such as the basal ganglia, and the GABA (gamma-aminobutyric acid)-producing Purkinje neurons. Early onset generalized dystonia is a dominantly inherited disorder caused by neuronal dysfunction without an apparent loss of neurons. Most cases of the dominantly inherited movement disorder, early onset torsion dystonia (DYT1) are caused by a mutant form of torsinA. The same single mutation (GAG deletion) causes most cases, and results in loss of a Glutamine (Glu) in the carboxy terminal region of torsinA (Δ302/303) (45). Relatively little is understood about the normal function of torsinA or the physiological effects of the codon deletion associated with most cases of disease. TorsinA is ATPase associated with diverse cellular activities (AAA) protein (46). It is apparently involved in membrane structure/movement and processing of proteins through the secretory pathway and is located predominantly in the lumen of the endoplasmic reticulum (ER) (47, 48). The mutant form of torsinA [torsinA(ΔE)] accumulates in the nuclear envelope (NE), leading to the formation of NE-derived membranous inclusions, known as spheroid bodies, which accumulate in the cytoplasm (49–51). Overexpression of wild-type torsinA in cultured cells results in a reticular distribution of immunoreactive protein that colocalizes with endoplasmic reticulum resident chaperones, while the dystonia-related mutant form accumulates within concentric membrane whorls and nuclear-associated membrane stacks (52).

Recent studies suggest that the mutant form of torsinA carrying the disease-linked mutation, torsinA(ΔE), acts through a dominant negative effect by recruiting wild-type torsinA into oligomeric structures in the nuclear envelope (53). Therefore, suppressing torsinA(ΔE) expression through RNAi could restore the normal function of torsinA(wt), representing a potentially effective therapy regardless of the biological role of torsinA (54). Gonzalez-Allegre et al. (55) generated shRNAs that mediate allele-specific suppression of torsinA(ΔE) and rescue cells from its dominant-negative effect, restoring the normal distribution of torsinA(wt). In addition, delivery of this shRNA by a recombinant feline immunodeficiency virus effectively silenced torsinA(ΔE) in a neural model of the disease. They established the feasibility of viral-mediated RNAi approach by demonstrating significant suppression of endogenous torsinA in mammalian neurons. Silencing of torsinA was achieved without triggering an interferon response. These results support the potential use of viral-mediated RNAi as a therapy for DYT1 dystonia and establish the basis for preclinical testing in animal models of the disease. Further, to model the neuronal involvement, adult rat primary sensory dorsal root ganglion neurons in culture were infected with lentivirus vectors expressing human wild-type or mutant torsinA (56). Expression of the mutant protein resulted in formation of torsinA-positive perinuclear inclusions. When the cells were coinfected with lentivirus vectors expressing the mutant torsinA message and a shRNA selectively targeting this message, inclusion formation was blocked. Vector-delivered siRNAs have the potential to decrease the adverse effects of this mutant protein in neurons without affecting wild-type protein.

A reduced rate of secretion is observed in primary fibroblasts from DYT1 patients expressing endogenous levels of torsinA and torsinA(ΔE) when compared with control fibroblasts expressing only torsinA (47). siRNA oligonucleotides have been used to downregulate the levels of torsinA or torsinA(ΔE) mRNA and protein by over 65% following transfection. Transfection of siRNA for torsinA mRNA in control fibroblasts expressing Gaussia luciferase reduced levels of luciferase secretion compared with the same cells nontransfected or transfected with a nonspecific siRNA. Transfection of siRNA selectively inhibiting torsinA(ΔE) mRNA in DYT fibroblasts increased luciferase secretion when compared with cells nontransfected or transfected with a nonspecific siRNA. Further, transduction of DYT1 cells with a lentivirus vector expressing torsinA, but not torsinB, also increased secretion (47), consistent with a role for torsinA as an ER chaperone affecting processing of proteins through the secretory pathway, and indicating that torsinA(ΔE) acts to inhibit this torsinA activity. The ability of allele-specific siRNA for torsinA(ΔE) to normalize secretory function in DYT1 patient cells supports its potential role as a therapeutic agent in early onset torsion dystonia.

4. Poly(Q)-Disorders

There are at least nine human neurodegenerative disorders that are caused by polyglutamine extension, i.e., by expansion of the CAG trinucleotide repeat. RNAi has recently been performed on several prevalent polyglutamine diseases, e.g., spinobulbar muscular atrophy, spinocerebellar ataxia, Machado–Joseph disease, alpha-synucleinopathies, and Huntington's disease.

Spinobulbar muscular atrophy is an X-linked motorneuron disease, related to CAG trinucleotide expansion in the first exon of the androgen receptor. It occurs at adulthood, and results in the loss of motorneurons in the lower spinal cord and the brain stem. Unaffected individuals have between 11 and 35 poly(Q) repeats as opposed to 38 and 62 repeats in patients (57, 58). This poly(Q) expansion results in intranuclear aggregates formation in tissues where the androgen receptor gene is normally expressed.

Drosophila S2 cells were engineered to express a portion of human androgen receptor gene with CAG tracts of 26, 43 or 106 repeats tagged by green fluorescent protein (GFP) (59). Cells carrying the CAG repeats of 43 and 106 developed GFP aggregates and aggresomes with 106 aggregates being formed much faster than 43 CAG repeats. Using RNAi directed against the androgen receptor mRNA, Caplen et al. (59) showed loss of aggregates by 80% in cotransfected S2 cells.

Spinocerebellar ataxia type 1 is another poly(Q) expansion disease related to CAG repeats of ataxin-1 and characterized by slowly progressive incoordination of gait, and often associated with poor coordination of hands, speech, and eye movements, frequently resulting in cerebellar atrophy. Xia et al. (60) have successfully suppressed these symptoms by RNAi in a mouse model of this disease (60). These studies show that RNAi interference could have considerable therapeutic potential in poly(Q) neurodegenerative disorders.

Machado–Joseph disease (MJD) is a related ataxia and a fatal, dominant neurodegenerative disorder. MJD results from polyglutamine repeat expansion in the MJD-1 gene, conferring a toxic gain of function to the ataxin-3 protein. Overexpressing ataxin-3 in the rat brain using lentiviral vectors enabled to study the disorder in defined brain regions (61). No neuropathological changes were observed upon wild-type ataxin-3 overexpression. Mutant ataxin-3 expression in striatum and cortex resulted in the accumulation of misfolded ataxin-3, and, within striatum, the loss of neuronal markers. Striatal pathology was confirmed by observation of ataxin-3 aggregates in MJD transgenic mice and substantial reduction of the dopamine and cAMP regulated phosphoprotein (DARPP-32) immunoreactivity and, in human striata, by ataxin-3 inclusions, immunoreactive for ubiquitin and alpha-synuclein (61). This interesting study demonstrates the use of LV encoding mutant ataxin-3 to produce a model of MJD.

Synucleinopathies are caused by aggregates of alpha-synuclein, due to polyglutamine extension of the mutated protein. The synuclein family includes three proteins, alpha-synuclein, beta-synuclein, and gamma-synuclein, sharing a common, highly conserved motif with similarity to the class-A2 lipid-binding domains of apolipoproteins. Their function is poorly known so far, but alpha-synuclein was found to be mutated in several families with autosomal dominant Parkinson's disease (62). Alpha-synuclein is primarily found in neural tissue, where it is seen mainly in presynaptic terminals. It is predominantly expressed in the neocortex, hippocampus, substantia nigra, thalamus, and cerebellum. It is predominantly a neuronal protein, but can also be found in glial cells (63). Alpha-synuclein overexpression and its toxic accumulation in neurons or glia are known to play key roles in the pathogenesis of Parkinson's disease and other related neurodegenerative synucleinopathies. It can aggregate to form insoluble fibrils in these pathological conditions characterized by "Lewy bodies", such as Parkinson's disease, dementia with Lewy bodies, and multiple system atrophy. In addition, an alpha-synuclein fragment, known as the nonAbeta component (NAC), is found in amyloid plaques in Alzheimer's disease.

In rare cases of familial forms of Parkinson's disease, there is a mutation in the gene coding for alpha-synuclein. Three point mutations have been identified thus far: A30P, E46K, and A53T. In addition, duplication and triplication of the gene appear to be the cause of Parkinson's disease in other lineages. Moreover, genetic variability of the alpha-synuclein gene promoter is associated with idiopathic Parkinson's disease.

To develop an animal model of PD, lentiviral vectors expressing different human or rat forms of alpha-synuclein were injected into the substantia nigra of rats (64). A selective loss of nigral dopaminergic neurons associated with a dopaminergic denervation of the striatum was observed in animals expressing either wild-type or mutant forms of human alpha-synuclein. This neuronal degeneration correlates with the appearance of abundant alpha-synuclein-positive inclusions and extensive neuritic pathology and recapitulates the essential neuropathological features of PD. The authors conclude that lentiviral-mediated genetic models may contribute to elucidate the mechanism of alpha-synuclein-induced cell death, and allow the screening of candidate therapeutic molecules (64).

Silencing of the human alpha-synuclein gene by vector-based RNAi is a promising therapeutic approach for synucleinopathies. Fontaine et al. (65) achieved 80% protein knockdown using siRNA targeted to endogenous alpha-synuclein and showed that alpha-synuclein suppression in vitro decreases dopamine transport and alters cellular dopamine homeostasis (65). Sapru et al. reported a 21-nucleotide sequence in the coding region of human alpha-synuclein that constitutes an effective target for robust silencing by RNAi (66). Using a dual cassette lentivirus that coexpresses an alpha-synuclein-targeting shRNA and EGFP (as a marker gene), effective silencing of endogenous human alpha-synuclein in vitro in the human dopaminergic cell line SH-SY5Y could be demonstrated. Similarly, the in vivo silencing of experimentally expressed human alpha-synuclein in rat brain has been successfully achieved (66). This may provide the tools for developing effective gene silencing therapeutics for synucleinopathies, including Parkinson's disease.

Alpha-synuclein inhibits tyrosine hydroxylase (TH), the rate-limiting enzyme in catecholamine synthesis, which is frequently used as a marker of dopaminergic neuronal loss in animal models of Parkinson's disease (PD). TH is activated by the phosphorylation of key seryl residues in the TH regulatory-domain.

Alpha-synuclein reduces TH phosphorylation, which then reduces dopamine synthesis (67, 68). Using lentiviral vectors expressing wt human alpha-synuclein, Alerte et al. (69) transduced dopaminergic neurons in the olfactory bulb of alpha-synuclein knockout mice and observed that cells bearing aggregated alpha-synuclein had significantly reduced total anti-TH immunoreactivity, but high phosphoserine-TH labeling, which suggests that aggregated alpha-synuclein is no longer able to inhibit TH (69).

Huntington's disease is an autosomal dominant hereditary brain disorder that is progressive and fatal, caused by expansion of a CAG trinucleotide repeat in exon-1 of the Huntingtin gene (70). The precise function of huntingtin is unknown, but in mice, complete deletion of the huntingtin gene is lethal. Mutations in this gene result in involuntary movement (chorea), cognitive impairment and psychiatric problems such as depression and anxiety (71, 72). This expansion elongates the N-terminal poly(Q) stretch of the protein, resulting in aggregation and the formation of neuronal intranuclear inclusions that causes an increase in the rate of neuronal cell death in select areas of the brain, affecting neurological functions (73, 74). As Huntington's disease is a monogenetic disorder conclusively linked to a single gene, researchers investigated using gene knockdown and lentiviral-mediated gene transfer of mutated huntingtin as a potential treatment (75, 76). siRNA therapy achieved a 60% reduction in expression of the mutated protein in a mouse model, and progression of the disease was stalled (77). However, using RNA interference to treat Huntington's disease could have unexpected effects unless knockdown of the unmutated huntingtin protein can be avoided, since full deletion of the normal protein is lethal. Also off-target effects of siRNA must be avoided (78).

A lentiviral vector expressing a mutant huntingtin protein (htt171-82Q) was used to generate a chronic model of Huntington's disease (HD) in rat primary striatal cultures (79). Primary striatal neurons expressing N-terminal fragments of mutant huntingtin in vitro (via lentiviral gene delivery) faithfully reproduced the gene expression changes seen in human patients and may serve as models of the disease (79–81). Expression of a wild-type fragment of huntingtin (htt171-18Q) caused only a small number of RNA changes. The disease-related signal in huntingtin171-82Q versus huntingtin171-18Q comparisons results in the differential detection of 20% of all mRNA probe sets (81). Transcriptomic effects of mutated huntingtin171 are time- and polyglutamine-length-dependent, and occur in parallel with other manifestations of polyglutamine toxicity over 4–8 weeks. Specific changes in lentivirus-mediated huntingtin171-82Q-expressing striatal cells accurately recapitulated those observed in human patients and included decreases in specific mRNAs (proenkephalin, regulator of G-protein signaling-4, dopamine D1 and D2 receptors, cannabinoid

CB1 receptor, and DARPP-32). This lentiviral-transfer study showed that induced dysregulation of the striatal transcriptome might be attributed to intrinsic effects of mutant huntingtin (81).

A new strategy based on lentiviral-mediated delivery of mutant huntingtin (htt) was used to create a genetic model of Huntington's disease (HD) in rats and to assess the relative contribution of polyglutamine (CAG) repeat size. Lentiviral vectors coding for the first 171, 853, and 1520 amino acids of wild-type (19 CAG) or mutant htt (44, 66, and 82 CAG) driven by either the phosphoglycerate kinase 1 (PGK) or the cytomegalovirus (CMV) promoters were injected in rat striatum (82). Lentiviral-mediated overexpression of mutated huntingtin expressing 82 polyglutamine residues protein led to proteolytic release of N-terminal huntingtin fragments, nuclear aggregation, and a striatal dysfunction, and accelerated the pathological process, as revealed by decrease of DARPP-32 staining but absence of the neuron-specific nuclear protein (NeuN) downregulation (83). Heat-shock proteins, which refold denatured proteins, also mitigate Huntington's disease mice models; as a matter of fact, based on lentiviral-mediated overexpression of a mutated huntingtin fragment, Perrin et al. (84) showed that heat-shock proteins 104 and 27 rescue striatal dysfunction in primary striatal neuronal cultures and in vivo, both chaperones significantly reduced mutated huntingtin-related loss of DARPP-32 expression in rat (84). Furthermore, the trophic factors CNTF (ciliary neurotrophic factor) and BDNF (brain-derived neurotrophic factor) are neuroprotective in acute Huntington's models and prevent striatal degeneration in a chronic model. Results demonstrated that both agents were neuroprotective without modifying inclusion formation (79). Lentiviral vectors expressing the human CNTF were injected in the striatum of wild-type and transgenic mice expressing full-length huntingtin with 72 CAG repeats; behavioral analysis showed increased locomotor activity in 5- to 6-month-old treated mice compared to wild-type animals (80). Interestingly, CNTF expression reduced the activity levels of mice expressing mutated huntingtin compared to control animals. Thus, sustained striatal expression of CNTF can be achieved with lentiviruses (80). By contrast, intrastriatal lentiviral vector transfer of GDNF, performed at 5 weeks of age, did not ameliorate neurological and behavioral impairments in the R6/2 transgenic mice model of Huntington's disease (85).

Another hypothesis for the disease suggested that dopamine, found in high concentrations in the striatum, may play a role in striatal cell death by increasing the toxicity of an N-terminal fragment of mutated huntingtin (Htt-171-82Q) (86, 87). Stimulation of the dopamine D2 receptor (DR2) with haloperidol protects striatal neurons from dysfunction induced by mutated huntingtin (88). Mutant huntingtin causes disruption of mitochondrial function by inhibiting expression of PGC-1α, i.e., a transcriptional co-activator that regulates several metabolic processes, including mitochondrial biogenesis and respiration, and lentiviral-mediated delivery of PGC-1αin the striatum provides neuroprotection in the transgenic mice models (89). Mitochondrial complex III-mediated modulation of huntingtin aggregates was observed in a neuronal progenitor RN33B cell line transduced by lentivirus carrying mutant huntingtin (90). Mitochondrial complex II protein (mCII) levels are reduced in striatum of patients, notably the 30-kDa iron-sulfur subunit and the 70-kDa FAD subunit, (86), indicating that this complex may be important for dopamine-mediated striatal cell death. Lentiviral-mediated expression of Htt171-82Q preferentially decreased the levels of these subunits in vitro and affected the dehydrogenase activity of the complex (86). Dopamine-induced down-regulation of mCII levels can be blocked by several dopamine D2 receptor antagonists (87). Lentiviral-mediated overexpression of mCII subunits abrogated the effects of dopamine, both by high dopamine concentrations alone and neuronal death induced by low dopamine concentrations together with Htt-171-82Q (87). Thus, lentiviral treatment enabled to explore a novel pathway that links dopamine signaling and regulation of mCII activity, and provides insights for a key role for oxidative energy in the disease.

5. Parkinson's Disease

Parkinson's disease is characterized by a progressive loss of midbrain dopamine neurons and the presence of cytoplasmic inclusions called Lewy bodies. Several approaches have been developed using lentiviral- mediated gene transfer technologies (91–94). Mutations in several genes have been linked to familiar Parkinson's, including alpha-synuclein and parkin. Parkin is a protein that functions within a multiprotein E3 ubiquitin ligase complex, catalyzing the covalent attachment of ubiquitin moieties onto substrate proteins, thus promoting degradation of specific substrates in cells. It is characterized by its ubiquitin-like motive at the N-terminus, and two RING fingers domains (RING I and RING II) and one "in-between RING finger motives" at the C-terminus (95). Parkin may protect neurons against α-synuclein toxicity, proteasomal dysfunction and kainate-induced excitotoxicity, and may play a role in controlling neurotransmitter trafficking at the presynaptic terminal and in calcium-dependent exocytosis. The protein carries out its activity by polyubiquitinating substrates through its lysine 48 residue for proteasome-dependent degradation (96), but it can also promote a proteasome-independent process by lysine 63-linked ubiquitination, which is a signal for ribosomal function, endocytosis of membrane proteins, protein sorting and trafficking (97, 98). The loss of parkin's E3-ligase activity leads to dopaminergic neuronal degeneration in early-onset autosomal recessive juvenile parkinsonism, suggesting a key role of parkin for dopamine neuron survival. Thus, its overexpression may provide a novel strategy for neuroprotection in the disease (99). In a rat lentiviral model of Parkinson's, animals overexpressing parkin showed significant reductions in alpha-synuclein-induced neuropathology, including preservation of tyrosine hydroxylase-positive cell bodies in the substantia nigra (100). The parkin-mediated neuroprotection was associated with an increase in hyperphosphorylated alpha-synuclein inclusions, suggesting a key role for parkin in the genesis of Lewy bodies.

As another approach, reactive oxygen species are considered to contribute to the pathogenesis of the disease. Therefore, antioxidative gene therapy strategies may be relevant as a therapeutical approach. Indeed, lentivirus-mediated expression of the antioxidant enzyme glutathione peroxidase provides small but significant neuroprotection against drug-induced toxicity (101), and could be suitable as a potential neuroprotective approach. In addition, an abnormal accumulation of cytosolic dopamine, resulting in reactive oxygen species and dopamine–quinone products, plays an important role in the selective degeneration of dopaminergic neurons of a part of brain called substantia nigra pars compacta in Parkinson's disease. The neuronal-specific vesicular monoamine transporter (VMAT2), responsible for uptake of dopamine into vesicles, has been shown to play a central role in both intracellular dopamine homeostasis and sequestration of dopaminergic neurotoxins. Lentivirus-mediated transfection of VMAT2 increases intracellular dopamine content, augments potassium-induced dopamine release and attenuates cell death, and enhances vesicular dopamine storage (102). The opposite was seen after downregulation of VMAT2 using virally delivered shRNAs (102).

Furthermore, growth factors such as glial cell line-derived neurotrophic factor (GDNF) have been shown to prevent neurodegeneration and promote regeneration in many animal models of Parkinson's disease (103). MPTP (1-methyl-4-phenyl-1,2,3,6-tetrahydropyridine) is a neurotoxin used to study the disease in animal models (mainly monkeys) that causes permanent symptoms of Parkinson's disease by killing certain neurons in the substantia nigra of the brain. MPTP is metabolized into the toxic cation 1-methyl-4-phenylpyridinium (MPP+) by the enzyme MAO-B of glial cells, which primarily kills dopamine-producing neurons in a part of the brain called the pars compacta of the substantia nigra. Infusion of recombinant human GDNF into the rat striatum markedly protected dopamine neurons against MPP+ induced toxicity (104), and thus GDNF is a good candidate agent for restoring functional reinnervation and/or neuroprotection of dopamine nigrostriatal system for the treatment of Parkinson's disease (94). Recently, recombinant lentiviral vectors were developed and used for effective GDNF gene delivery (3, 92–94, 105). In vivo gene delivery of lentiviral vector–GDNF has been widely investigated (106), and viral delivery of GDNF currently represents one of the most promising neuroprotective strategies for Parkinson's Disease (100). However, one of the important unresolved issues for this strategy is the threshold number of dopamine (DA) nigral neurons and/or of striatal DA terminals necessary for optimal benefit from GDNF therapy (94). Intracerebral injection of lentiviral vector–GDNF into 6-hydroxydopamine (6-OHDA) lesioned rats (107–110) and MPTP-treated monkeys (105, 111) resulted in persistent protection of the nigrostriatal DA pathway and recovery of motor function (93, 108, 109, 112, 113), thereby strongly supporting their application in the clinic. Intrastriatal neurotrophic effects of long-term GDNF delivery using a lentiviral vector restored complete striatal dopamine innervation in the previously denervated area, and this was associated with significant behavioral improvements in a model of early Parkinson's disease (94). Apomorphine-induced rotation was significantly decreased in the lentiviral vector-GDNF-injected group compared with control animals (3) and GDNF efficiently protected 69.5% of the tyrosine hydroxylase-positive cells in the substantia nigra against 6-hydroxydopamine-induced toxicity. Lentiviral vector-GDNF treatment induced substantial sprouting from surviving DA axons, in agreement with previous work (75, 105, 114). Interestingly, aberrant sprouting was not observed, and consequently, negative functional effects, such as those presented by animals with extensive sprouting in or adjacent to the substantia nigra (SN) (75, 109), was not observed. However, a rat model has been used which develops a progressive and selective loss of dopamine neurons associated with the appearance of alpha-synuclein containing inclusions, thus recapitulating the major hallmarks of Parkinson's disease. In this model, a lentiviral vector coding for GDNF (lenti-GDNF) was tested for its ability to prevent nigral dopaminergic degeneration associated with the lentiviral-mediated expression of the A30P mutant human alpha-synuclein (lenti-A30P). When injected in the substantia nigra 2 weeks before nigral administration of lenti-A30P, it induced robust expression of GDNF but did not prevent the alpha-synuclein-induced dopaminergic neurodegeneration, indicating that sustained GDNF treatment cannot modulate the cellular toxicity related to abnormal protein accumulation (115). Recently, however, in an encouraging primate study, Kordower et al. (105) injected lentiviral vector-GDNF into striatum and substantia nigra 1 week after an intracarotid injection of the MPTP toxin, selectively damaging the dopamine system (103). In 3 months, motor deficits were reversed and loss of TH-positive neurons in the nigra had been prevented, while TH-positive projections to the striatum were preserved. Thus, if GDNF is delivered at the lesion site before neurodegeneration progresses too far, lentiviral vector-mediated GDNF treatments displays impressive efficiency and induces functional reinnervation, in both rodents and monkeys, which may provide a basis for new clinical trials (103).

Finally, as an alternative, human neural progenitor cells hold great potential as an ex vivo system for delivery of therapeutic proteins to the central nervous system. When cultured as aggregates (neurospheres), they are capable of significant in vitro expansion. Methods have been developed for lentiviral vector-mediated gene delivery into human neural progenitor cells that maintain their differentiation and proliferative properties, and result in long-term, stable expression, and these cells have been successfully transplanted into a rodent model of Parkinson's disease, inducing stable neuroprotection (116). Using a similar approach, Ebert et al. (117) produced human neural progenitor cells releasing either GDNF or Insulin-like growth factor 1 (IGF-1) and transplanted them into a rat model of PD. IGF-1, like GDNF, is also known to have neuroprotective effects in a number of disease models, but has not been extensively studied in models of PD. Progenitor cells secreting either GDNF or IGF-1 significantly reduced amphetamine-induced rotational asymmetry and dopamine neuron loss (117). Interestingly, GDNF, but not IGF-1, was able to protect or regenerate TH-positive fibers in the striatum, whereas in contrast, IGF-1, but not GDNF, significantly increased the overall survival of progenitor cells both in vitro and following transplantation (117). In areas of GDNF delivery, increased TH- and VMAT2 (vesicular monoamine transporter 2)-positive fibers were also observed, which may regulate the sensitivity of rat dopaminergic neurons to disturbed cytosolic dopamine levels.

Together, these studies show that lentiviral vectors constitute a powerful gene delivery system for the screening of therapeutic molecules in mouse models of Parkinson's disease (3, 91, 118). Due to their impressive efficiency, time may now be ripe to explore these vector systems as tools for neuroprotective treatments in patients with Parkinson's disease (113).

6. Alzheimer's Disease

Alzheimer's disease (AD) is associated with the accumulation of insoluble plaques of amyloid protein (APP) in the central nervous system, as well as intracellular microfibrillar tangles, along with loss of cholinergic neurons projecting from the basal forebrain to the hippocampus and amygdala. Plaque deposition is a result of overproduction of amyloid beta protein (Abeta) through the activity of gamma-secretase, of which the presenilins are thought to be a component. Early onset familial Alzheimer's, for example, is associated with mutations in presenilin genes PS1 and PS2 which lead to over secretion of Abeta. RNAi knockdown of PSN results in a blockage of γ-secretase activity (119). Also siRNAs targeting the β-secretase, BACE1, were shown to reduce APP production in mouse cortical neurons, offering a potential therapeutic approach for AD (120). Lentivirus vectors targeting selective genes involved in the disease have been very powerful in better clarifying their functions in the disorder.

Abeta is derived from APP by sequential cleavages of β- and γ-secretases. The molecules responsible for these proteolytic activities have been identified (121). The presenilin 1 and 2 are γ-secretase, and the identity of β-secretase has been shown to be the novel transmembrane aspartic protease, β-site APP-cleaving enzyme 1 (BACE1; also called Asp2 and memapsin 2). BACE2, a protease homologous to BACE1, was also identified, and together the two enzymes define a new family of transmembrane aspartic proteases (121). β-site APP cleaving enzyme 1 (BACE1) is the major β-secretase in vivo. Burton (122) designed a number of siRNA target sequences and cloned them into lentivirus vectors and showed that RNA inhibition of β-secretase reverts Alzheimer's disease in mice (122). Sierant et al. (123) designed lentivirus vector-encoded siRNAs for efficient knockdown of overexpressed and endogenous BACE1 in a rat brain and also the allele-specific siRNAs to silence the mutant presenilin 1 (L392V PS-1), the main component of γ-secretase. This results in strongly reducing the level of β-amyloid accumulation in the brain. These tools could be beneficial for potential therapeutic approach for treatment of the disease (123). In another study, Singer et al. (124) targeted BACE1 with siRNAs. Lowering BACE1 levels using these lentiviral vectors expressing siRNAs targeting BACE1 reduced amyloid production and the neurodegenerative and behavioral deficits in APP transgenic mice, an approach that could have potential therapeutic value for treatment of the disease (124). β-site APP cleaving enzyme 2 (BACE2) is the homologue of BACE1 and cleaves APP at a novel θ site downstream of the α site, abolishing Abeta production (125). Overexpression of BACE2 by lentiviral vectors markedly reduced Abeta production in primary neurons derived from Swedish mutant APP transgenic mice. This suggests that therapeutic interventions that potentiate BACE2 may prevent AD pathogenesis (125). Besides, El-Amouri et al. (126) targeted neprilysin (NEP), the dominant Abeta peptide-degrading enzyme in the brain. NEP becomes inactivated and downregulated during both the early stages of AD and aging. Lentiviral vector-mediated overexpression of human NEP for 4 months in young APP/ΔPS1 double-transgenic mice resulted in reduction in Abeta peptide levels, attenuation of amyloid load, oxidative stress, and inflammation, and improved spatial orientation. The overall reduction in amyloidosis and associated pathogenetic changes in the brain resulted in decreased memory impairment by approximately 50%. Thus, restoring NEP levels in the brain at the early stages of AD is an effective strategy to prevent or attenuate disease progression. The cysteine protease cathepsin B (CatB) is also associated with amyloid plaques in AD brains and reduces levels of Abeta peptides, especially the aggregation-prone species. Lentiviral vector-mediated expression of CatB in aged hAPP mice reduced preexisting amyloid deposits (127), even thioflavin S-positive plaques. Under cell-free conditions, CatB effectively cleaved Abeta1-42, generating C-terminally truncated Abeta peptides that are less amyloidogenic, and increasing CatB activity could counteract the neuropathology of this disease (127, 128).

Apolipoprotein E (apoE) alleles are important genetic risk factors for AD, which influence brain amyloid-β peptide (Abeta) and amyloid burden. Direct intracerebral administration of lentiviral vectors expressing the three common human apoE isoforms differentially alters hippocampal Abeta and amyloid burden in mouse model of AD (129). Expression of apoE4 in the absence of mouse apoE increases hippocampal Abeta (1-42) levels and amyloid burden. By contrast, expression of apoE2, even in the presence of mouse apoE, markedly reduces hippocampal Abeta burden. Thus, gene delivery of apoE2 may reduce the development of neuritic plaques (129).

Recently, Kim et al. showed that SIRT1, a human homologue of SIR2 (a gene which promotes longevity in a variety of organisms), is upregulated in mouse models for Alzheimer's disease (130). Injection of SIRT1-expressing lentiviral vectors in the hippocampus of the inducible p25 transgenic mice (a model of AD) conferred significant protection against neurodegeneration, reduced neurodegeneration in the hippocampus, prevented learning impairment, and decreased the acetylation of the known SIRT1 substrates PGC-1alpha and p53 (130). Neurodegeneration induced by pathogenic proteins depends on contributions from surrounding glia. In Alzheimer's disease, NF-κB signaling in microglia is critically involved in neuronal death induced by Abeta peptides. Stimulation of microglia with Abeta increases acetylation of RelA/p65, which regulates the NF-κB pathway. Overexpression of SIRT1 deacetylase and the addition of the SIRT1 agonist resveratrol markedly reduces NF-κB signaling stimulated by Abeta, and has strong neuroprotective effects, which support a glial loop hypothesis by demonstrating a critical role for microglial NF-κB signaling in Abeta-dependent neurodegeneration (131). Microglia, the immune cells of the brain, are activated in the brain of Alzheimer's disease patients, and express the innate immune receptor toll-like receptor 2 (TLR2). Richard et al. (132) generated triple transgenic mice deficient in TLR2 and harboring a mutant human presenelin 1 and a chimeric mouse/human amyloid precursor protein (APP). TLR2 deficiency accelerated spatial and contextual memory impairments, which correlated with increased levels of Abeta (1-42) and transforming growth factor β1 in the brain. Expression levels of N-Methyl-d-Aspartate (NMDA) receptors 1 and 2A were also lower in the hippocampus of APP-TLR2(−/−) mice. Gene therapy in cells of the bone marrow using lentiviral vectors expressing TLR2 rescued the cognitive impairment of APP-TLR2(−/−) mice and restored the memory consolidation process disrupted by TLR2 deficiency in APP mice (132).

7. Drug Addiction

Drug addiction is a major public health issue. It is typically a multigenetic brain disorder, implying combined changes of expression of several hundred genes. Addictive drugs (cocaine and amphetamines), depressants (ethanol) and opiate narcotics (heroin and morphine) are powerful reinforcers and produce their rewarding effects of euphoria or pleasure through an interaction with the mesolimbic dopamine system (133). The dopamine D_3 receptor (D_3R) is an important pharmacotherapeutic target for its potential role in psychiatric disorders and drug dependence. Lentiviral vector-mediated manipulations of D_3R levels, using local injection in the Accumbens of combinations of either D_3R-expressing and/or siRNAs-expressing lentiviruses induced strong behavioral changes in the locomotor stimulant effects of cocaine (134). Suppression of D_3R expression increased locomotor stimulant effects, whereas its overexpression with lentiviral vector-expressing D_3R drastically reduced them. The latter effects could be reversed when animals were fed doxycycline that prevented ectopic overexpression of D_3R in the Accumbens (134). In the mesolimbic dopaminergic pathway, dopamine is removed by its transporter DAT. lentiviral vector-mediated overexpression of DAT in the nucleus Accumbens induced a 35% decrease in locomotor activity, which could be abolished when the same animal was fed doxycycline, whereas local inhibition of DAT in the this brain area, using lentiviral vectors expressing siRNAs targeted against DAT, resulted in significant hyperlocomotion activity (72% increase over controls) (135). α-Synuclein binds to the dopamine and serotonin transporters (DAT and SERT), affecting their activities and their recruitment to the plasma membrane, an effect that can be modulated by γ-synuclein expression (136, 137). Local manipulation of either α- or γ-synuclein expression by means of lentiviral vectors expressing either genes or siRNAs targeted against them, strongly affected cocaine-induced behavioral effects and γ-synuclein modulates cocaine-induced reinforcing and incentive effects (138, 139).

Strong molecular adaptations and plasticity within this mesolimboc dopaminergic pathway upon administration of drugs underlie a complex rewiring of neural circuitry that results in the behaviors associated with addiction (133, 140). Little is known about the specific targets involved in this neuroadaptation process, but it has been suggested that cocaine and other drugs of abuse may alter the morphology of neuronal dendrites and spines, the primary site of excitatory synapses in the brain, by means of inducing expression changes of surface molecules (133, 141). Complex changes in expression of a number of surface axon guidance molecules have been observed upon cocaine administration, which may underlie important neuroplastic changes in the reward- and memory-related brain centers after drug action (142). Local expression changes of these cues may mediate plasticity and cytoskeleton rearrangement through mechanisms similar to synaptic targeting during development. In addition, strong induction of a surface tetraspanin protein involved in cell adhesion, CD81, has been described (143–145). Tetraspanins are scaffold proteins that integrate receptor signaling and form complexes with integrins, to mediate axon growth induced by numerous extracellular matrix proteins. Using a regulatable lentiviral vector bearing the rat CD81 gene under the control of a tetracycline-inducible system, Bahi et al. (146) showed that local overexpression of CD81 in this pathway, by means of doxycycline-regulatable lentiviral-mediated gene transfer into the Nucleus Accumbens, induced a fourfold to fivefold increase in locomotor activity that can be reversed to normal in the same animal fed doxycycline (144, 146). Furthermore, several lentiviral vectors expressing siRNA targeted against CD81 have been designed and CD81 gene silencing was assessed in vivo upon local, stereotaxic injection into the ventral tegmental area or the Nucleus Accumbens (147). Expression of siRNA targeted against CD81 decreased CD81-mediated locomotor activity almost back to basal levels, and could be regulated by very local and specific gene knockdown, a clear indication that drug-mediated behavioral activity is strongly modulated by this tetraspannin (147).

A similar approach was used to assess the role of other genes in drug addiction. Among these, the plasminogen system is particularly intriguing. Extracellular proteolytic processes play a key function in regulating synaptic structure and function, and CNS plasticity. Among these, plasminogen activators are important mediators of extracellular metabolism, involved in remodeling events during development and regeneration in the nervous system. The generation of plasmin from its inactive precursor plasminogen, is mediated by serine enzymes known as tissue-type plasminogen activator (tPA) and urokinase (uPA), and contributes to the turnover of the extracellular matrix in the CNS. Urokinase-type plasminogen activator (uPA) exerts a variety of functions during development, and is involved in learning and memory. Several tetraspanins drastically affect plasmin activity and prevent uPA binding to its receptor uPAR by promoting stable association of uPAR with specific integrins (e.g., $\alpha5\beta1$), redistributing uPAR to focal adhesion. A screening for genes induced after drug treatment showed that urokinase plasminogen-type activator (uPA) was strongly regulated by cocaine in several paradigms of drug administration (148). Lentiviral vector-mediated injection of a doxycycline-regulated uPA expression cassette, (or of its mutated form) into specific rat brain areas (Hippocampus, Nucleus Accumbens and Ventral Tegmental Area) showed a dramatic, doxycycline-dependent, 10- to 12.3-fold increase in locomotor activity after cocaine administration (148). Behavioral effects were completely abolished when uPA expression was turned off with doxycycline or when the active site of the protease was point-mutated and used as a dominant negative (148, 149). Lentiviral vector-mediated overexpression of uPA in the Ventral Tegmental Area induces doxycycline-dependent expression of its receptor, uPAR, but not its inhibitor, plasminogen activator inhibitor-1 (PAI-1) (149). Local injection of uPA-specific siRNAs expressing lentiviral vectors into the ventral tegmental area suppressed cocaine-induced behavioral changes associated with uPA expression (149, 150) Lentivirus vector-mediated overexpression of uPA in the Nucleus Accumbens shell significantly augments cocaine-induced place preference, whereas its inhibition abolished the acquisition but not the expression of cocaine-induced place preference (149, 150). Furthermore, while this overexpression did not affect the ability of preference to be extinguished, reinstatement with a low dose of cocaine produced significantly greater preference to the cocaine-associated context. Thus, once place preference had been established, and the preference extinguished, reinstatement induced by a priming dose of cocaine is facilitated by uPA. Furthermore, lentiviral vector-mediated overexpression of uPA induces cocaine-, but not amphetamine-mediated behavioral changes, whereas, in contrast, tPA-overexpressing animals displayed greater place preference when administered amphetamine or morphine compared to uPA-overexpressing animals (149–151). The behavioral effects are suppressed when tPA has been silenced using specific siRNAs-expressing lentiviral vectors. Suppression of endogenous tPA expression in animals treated with siRNA-expressing lentiviral vectors also fully suppresses place preference and these animals appear to avoid the drug-associated box. In addition, lentiviral vectors-mediated tPA overexpression delays extinction, but priming with low doses of amphetamine (not cocaine) reinstates place preference even after full extinction. Thus, tPA plays an important role in acquisition of amphetamine-induced conditioned place preference but its role in place preference-expression does not seem important (151, 152). Together, these data indicate that tPA and uPA may induce distinct behaviors, which may be interpreted according to their differential pattern of activation and downstream targets. Thus, modulation of the plasminogen system in the brain might be a potential target against drugs of abuse (153).

Furthermore, the neurotrophin BDNF (brain-derived neurotrophic factor) is also involved in cocaine reward and relapse. BDNF is involved in the survival and function of midbrain dopamine neurons, mediated by its TrkB receptor. BDNF and TrkB transcripts are widely expressed in the mesolimbic pathway, including the Nucleus Accumbens and the Ventral Tegmentum Area. Lentiviral vectors expressing either BDNF or TrkB enhance drug-induced locomotor activity and induce sensitization in rats (135). Upon local treatment, animals display enhanced cocaine-induced place preference, delayed place preference-extinction and increased reinstatement. Lentiviral vector-mediated expression of TrkT1 (truncated form of TrkB, acting as a dominant negative) inhibits these behavioral changes. This inhibition is also observed when rats are fed doxycycline (blocking local, lentivirus-mediated gene expression) or when injected with siRNAs-expressing lentiviruses against TrkB (135). This approach enabled to show that BDNF and TrkB-induced place preference takes place during the learning period (conditioning), whereas extinction leads to the loss of place preference. Extinction is delayed when rats are injected BDNF- or TrkB-expressing lentiviral vectors, whereas priming injections of cocaine reinstates it.

8. Attention-Deficit Hyperactivity Disorder

Attention-Deficit Hyperactivity Disorder (ADHD) is a neurobehavioral developmental disorder characterized by a persistent pattern of impulsiveness and inattention, with or without a component of hyperactivity (154). According to the dominant model, ADHD is viewed as an executive dysfunction (155–157), but alternative accounts present ADHD as a motivational dysfunction (158), arising from altered reward processes within fronto-striatal circuits (159, 160). Comorbidity to ADHD and other impulse-control disorders is pathological gambling as well as common, especially compulsive buying and compulsive sexual behavior (161). The disorder is highly heritable and arises from a combination of various genes, many of which affect dopamine transporters. Psychogenetic studies have revealed that, among several genes involved in DAergic neurotransmission within the reward system, one of the most significant being the dopamine transporter (DAT) (162). Besides, candidate genes include the dopamine receptor D4 (163), dopamine β-hydroxylase (164), monoamine oxidase A, catecholamine-methyl transferase, serotonin transporter promoter, and serotonine receptors 5-HT2A and 5-HT1B (165, 166).

Adriani et al. (167) tested the role of striatal dopamine function in ADHD using lentiviral vectors driving the expression of DAT or siRNAs targeted against DAT mRNA in the Nucleus Accumbens of rats. These tools enabled us to evaluate behavioral changes, associated with very local DAT overexpression or suppression (167). Accumbal DAT suppression was expected to enhance tonic DA transmission, compared to controls, whereas its overexpression would drastically reduce synaptic DA levels. To probe the resulting behavioral phenotype, animals were tested for several motivational endpoints such as seeking for novelty, intolerance to delay, and temptation to gamble (168, 169).

Whereas all knockout mice models have the major disadvantage that the gene is absent all over the CNS, lentiviral vector-mediated DAT manipulation is very local and only initiated in mature animals. Rats were followed both for socioemotional profiles and for risk-seeking propensity. Elevated anxiety and affiliation towards a stranger emerged upon silencing DAT in the nucleus accumbens, in rats stereotaxically injected with lentiviral vector-expressing shRNA targeted to DAT. Interestingly, using a doxycycline-regulatable lentiviral vector overexpressing DAT, it was showed that levels of playful social interaction were markedly reduced compared to controls. These rats displayed a marked "gambling" profile (i.e., preference for a large/uncertain over a small/sure reward), which disappeared upon doxycycline-induced switch-off on DAT enhancer, and consistently reappeared with doxycycline removal (167). This powerful tool allowed to reproduce an integrated profile in a regulatable way and shows that lentiviral vectors are suitable tools for animal models to reproduce psychiatrically relevant symptoms.

9. Other Disorders

In vivo gene delivery, by means of lentiviral vector mediated gene transfer, has been used for assessing gene function of many other neuronal pathologies. In many cases, lentiviral vector-mediated in vivo delivery of shRNAs targeting a major molecular candidate has been attempted also.

RETT Syndrome. The methyl-CpG-binding protein-2 gene (Mecp2) is the causal gene of the neurodevelopmental disorder, Rett Syndrome. A rat model with a downregulated Mecp2 was established using a recombinant lentiviral vector expressing small hairpin RNA of the rat Mecp2 gene (170). Four recombinant vectors were constructed by inserting sequences of small hairpin RNA targeting the rat Mecp2 gene. Rats were given intraventricular injections. Mecp2 mRNA was lower in the hippocampus and cerebral cortex relative to control groups. Although no typical Rett-like symptoms were observed, the neonatal rats injected with recombinant lentivirus displayed some transient neurobehavioral abnormalities during early development (170). BDNF mRNA expression decreased in the hippocampus, supporting the hypothesis that BDNF may be a target gene of MeCP2 in the CNS (170). Furthermore, by means of a lentiviral vector expressing Cre recombinase, Nelson et al. (171) showed that loss of MeCP2 function after neurodevelopment and synaptogenesis was sufficient to mimic the decrease in the frequency of spontaneous excitatory synaptic transmission seen in constitutive MeCP2 KO neurons (171). Taken together, these results suggest a role for MeCP2 in control of excitatory presynaptic function through regulation of gene expression.

Amyotrophic Lateral Sclerosis. Amyotrophic lateral sclerosis (ALS) is a fatal neurodegenerative disease resulting in the selective death of motor neurons in the brain and spinal cord. Mutations in Cu/Zn superoxide dismutase (encoded by *SOD1*) are one of the causes of familial amyotrophic lateral sclerosis that leads to progressive death of motoneurons through a gain-of-function mechanism. RNA interference (RNAi) mediated by lentiviral vectors allows for long-term reduction in gene expression and represents an attractive therapeutic approach for genetic diseases characterized by acquired toxic properties (172). Lentiviral vectors mediating expression of RNAi molecules, specifically targeting the human *SOD1* gene (*SOD1*), have been designed and injected into various muscle groups of mice engineered to overexpress a mutated form of human *SOD1* (*SOD1*G93A), which resulted in an efficient and specific reduction of *SOD1* expression and improved survival of vulnerable motor neurons in the brainstem and spinal cord (172). Furthermore, in *SOD1*G93A transgenic mice (a model for familial ALS), intraspinal injection of a lentiviral vector that produces RNAi-mediated silencing of *SOD1* substantially retards both the onset and the progression rate of the disease (172, 173).

Scrapie. Prion diseases are fatal neurodegenerative diseases characterized by the accumulation of PrPSc, the infectious and protease-resistant form of the cellular prion protein (PrPC). Pfeiffer et al. (174) generated lentivectors expressing PrPC-specific shRNAs that efficiently silenced expression of the prion protein gene (*Prnp*) in primary neuronal cells. Treatment of scrapie-infected neuronal cells with these lentivectors resulted in an efficient and stable suppression of PrPSc accumulation. After intracranial injection, lentiviral shRNA reduced PrPC expression in transgenic mice carrying multiple copies of *Prnp* (174).

Glioma. Glioma cells are characterized by their invasiveness and resistance against conventional therapeutics. Telomerase activity has been suggested to be an important target for glioma treatment. Zhao et al. (175) assessed the anticancer effects of lentiviral vector-mediated siRNA knockdown of the human telomerase reverse transcriptase (hTERT) in U87MG human glioblastoma cells and in vivo. Injection of lentiviral vectors significantly inhibited the growth of preestablished macroscopic xenograft tumors. The in vivo glioma growth inhibition effect coincided with no detectable telomere length changes (2, 175). Thus, efficient knockdown of hTERT can inhibit glioma cell proliferation and migration prior to its effect on telomere length.

Spinal cord injury. Injuries to the adult mammalian spinal cord often lead to severe damage to both ascending (sensory) pathways and descending (motor) nerve pathways without the perspective of complete functional recovery. Neurotrophins have emerged as promising molecules to augment neuroprotection and neuronal regeneration. Recently, lentiviral vector-mediated gene transfer of neurotrophins has been tested successfully in order to promote regeneration of the injured spinal cord (176). Furthermore, a conditioning lesion to peripheral axons of primary sensory neurons accelerates regeneration of their central axons in vivo or neurite outgrowth. Neuropoietic cytokines are also involved in this in regenerative conditioning (177). Delivery through a lentiviral vector of ciliary neurotrophic factor to the appropriate dorsal root ganglion in rats effectively mimics the conditioning effect of peripheral nerve injury on the regeneration of dorsal spinal nerve root axons (177).

Other uses. Finally, lentiviral vector-mediated gene transfer has further great potential and applications. For example, human neural progenitor cells hold great potential as an ex vivo system for delivery of therapeutic proteins to the central nervous system, because, as aggregates (neurospheres), they are capable of significant expansion. Capowski et al. (115) developed lentiviral vector-mediated genetic modification of human neural progenitor cells for ex vivo gene therapy that maintains the differentiation and proliferative properties of neurosphere cultures, while minimizing the amount of viral vector used and controlling the number of insertion sites per population. This method results in long-term, stable expression even after differentiation of progenitor cells to neurons and astrocytes and may have great therapeutic potential. In another approach, Dittgen et al. (178) established a method for genetic manipulation and subsequent phenotypic analysis of individual cortical neurons in vivo. Lentiviral vectors were prepared for neuron-specific gene delivery from either calcium-calmodulin-dependent protein kinase II promoter or from Synapsin I promoter, optionally in combination with gene knockdown by means of U6 promoter-driven expression of short-interfering RNAs (178).

These tools may be ideally suited for analysis of gene functions in individual neurons in the intact brain. Similarly, Kameda developed lentiviral vectors for targeting green fluorescent protein specifically to dendritic membrane in central neurons (179). Using a lentiviral vector with a neuron-specific promoter, GFP with a palmitoylation (palGFP) or myristoylation/palmitoylation site (myrGFP) was expressed in rat brain to target dendritic membranes efficiently. myrGFP, tagged with the C-terminal cytoplasmic domain of low density lipoprotein receptor, proved to be an excellent synthetic protein for dendritic visualization, and may be a useful tool for the morphological analysis of neuronal circuits (179). In an even more specific approach, Santamaria et al. developed a lentiviral vector producing shRNA targeting choline acetyl-transferase (ChAT) mRNAs and used this tool to transduce cholinergic neurons in vivo (180), resulting in a both strong and specific reduction of ChAT expression in the cholinergic neurons of the medial septum in adult rats without affecting the expression of the vesicular acetylcholine transporter. This lentiviral vector is thus a powerful tool for specific inactivation of cholinergic neurotransmission and can therefore be used to study the role of cholinergic nuclei in the brain.

10. Conclusions

RNA interference has now developed into a very powerful approach toward therapy for neural disorders (181), and many gene delivery strategies for RNAi knockdown of neuronal genes have been tested. In this review, I have shown the applications of lentiviral vector-mediated gene transfer in the investigation of a large variety of brain disorders. As a matter of fact, lentiviral vectors are increasingly used for gene delivery to neurons and in experimental models of brain disorders. Their use in gene delivery has received greatest attention and the use of neuron-specific promoters, where introduced, enables cell-specific gene transfer in a complex system such as the brain. The development of self-inactivating lentiviral vectors enables also to very locally manipulate the expression of a specific gene, in a very specific and limited neuronal pathway within the brain. In addition, the development of lentiviral systems has facilitated the exogenous manipulation of RNAi in hard to transfect, postmitotic cells such as neurons. Lentiviral vectors with their high tropism for neurons offer one of the most attractive options for delivering shRNAs to the CNS, as well as to cultured neurons and neuronal cell lines in vitro. This approach has been successfully used to functionally silence genes in primary mammalian cells, stem cells, and more importantly in transgenic mice, resulting in persistent knockdown of gene expression.

The use of lentiviral vectors for stable gene silencing in brain will soon prove a powerful aid to probe gene function in vivo and for gene therapy of diseases of the central nervous system. Lentiviral vector-mediated RNAi experiments in cultured mammalian neurons can be designed to address a wide variety of biological questions or to test potential therapeutic hairpins before moving to treatment trials in vivo. Indeed, potent silencing of gene expression in vitro and in vivo by lentiviral vector-based RNAi provide the tools for developing effective gene silencing therapeutics for many major brain disorders (182). Experimental gene therapy approaches under development using this technology are well advanced today, particularly for motor disorders such as Parkinson's disease and related synucleinopathies, Huntington's disease, dystonia and muscular dystrophy, as I have shown in this review. Furthermore, lentivirus gene transfer has been an invaluable tool recently for evaluation of gene function in highly complex behavioral disorders such as drug addiction and attention-deficit hyperactivity disorder or in learning and cognition. Many studies also include lentiviral vector-mediated delivery of neurotrophic factor genes, antiapoptotic genes and genes that modulate neurotransmission. Clearly, this technology will greatly help at a deeper understanding of these complex disorders in future studies.

References

1. Aebischer, P., and Ridet, J. L. (2001) Recombinant proteins for neurodegenerative diseases: the delivery issue. *Trends Neurosci* **24**, 533–40.

2. Zhao, C., Strappe, P. M., Lever, A. M. L., and Franklin, R. J. M. (2003) Lentiviral vectors for gene delivery to normal and demyelinated white matter. *Glia* **42**, 59–67.

3. Bensadoun, J. C., Deglon, N., Tseng, J. L., Ridet, J. L., Zurn, A. D., and Aebischer, P. (2000) Lentiviral vectors as a gene delivery system in the mouse midbrain: cellular and behavioral improvements in a 6-OHDA model of Parkinson's disease using GDNF. *Exp Neurol* **164**, 15–24.

4. Janas, J., Skowronski, J., Van Aelst, L. (2006) Lentiviral delivery of RNAi in hippocampal neurons. *Methods Enzymol* **406**, 593–605.

5. Gascón, S., Paez-Gomez, J. A., Díaz-Guerra, M., Scheiffele, P., and Scholl, F. G. (2008) Dual-promoter lentiviral vectors for constitutive and regulated gene expression in neurons. *J Neurosci Methods* **168**, 104–12.

6. Dillon, A. K., Fujita, S. C., Matise, M. P., Jarjour, A. A., Kennedy, T. E., Kollmus, H., Arnold, H. H., Weiner, J. A., Sanes, J. R., Kaprielian, Z. (2005) Molecular control of spinal accessory motor neuron/axon development in the mouse spinal cord. *J Neurosci* **25**, 10119–30.

7. Dykxhoorn, D. M., Novina, C. D., Sharp, P. A. (2003) Killing the messenger: short RNAs that silence gene expression. *Nat Rev Mol Cell Biol* **4**, 457–67.

8. Elbashir, S. M., Lendeckel, W., and Tuschl, T. (2001) RNA interference is mediated by 21- and 22-nucleotide RNAs. *Genes Dev* **15**, 188–200.

9. Hannon, G. J., and Rossi, J. J. (2004) Unlocking the potential of the human genome with RNA interference. *Nature* **431**, 371–8.

10. Scherr, M., Battmer, K., Ganser, A., and Eder, M. (2003) Modulation of gene expression by lentiviral-mediated delivery of small interfering RNA. *Cell Cycle* **2**, 251–7.

11. Buckingham, S. D., Esmaeili, B., Wood, M., and Sattelle, D. B. (2004) RNA interference: from model organisms towards therapy for neural and neuromuscular disorders. *Hum Mol Genet* **13**, 275–288.

12. Van den Haute, C., Eggermont, K., Nuttin, B., Debyser, Z., and Baekelandt, V. (2003) Lentiviral vector-mediated delivery of short hairpin RNA results in persistent knockdown

of gene expression in mouse brain. *Hum Gene Ther* **14**, 1799–807.

13. Rubinson, D. A., Dillon, C. P., Kwiatkowski, A. V., Sievers, C., Yang, L., Kopinja, J., Rooney, D. L., Ihrig, M. M., McManus, M. T., Gertler, F. B. Scott, M. L., Van Parijs, L. (2003) A lentivirus-based system to functionally silence genes in primary mammalian cells, stem cells and transgenic mice by RNA interference. *Nat Genet* **33**, 401–6.

14. Stewart, S. A., Dykxhoorn, D. M., Palliser, D., Mizuno, H., Yu, E. Y., An, D. S., Sabatini, D. M., Chen, I. S., Hahn, W. C., Sharp, P. A., Weinberg, R. A., and Novina, C. D. (2003) Lentivirus-delivered stable gene silencing by RNAi in primary cells. *RNA* **9**, 493–501.

15. Ventura, A., Meissner, A., Dillon, C. P., McManus, M., Sharp, P. A., Van Parijs, L., Jaenisch, R., and Jacks, T. (2004) Cre-lox-regulated conditional RNA interference from transgenes. *Proc Natl Acad Sci U S A* **101**, 10380–5.

16. Kunath, T., Gish, G., Lickert, H., Jones, N., Pawson, T., and Rossant, J. (2003) Transgenic RNA interference in ES cell-derived embryos recapitulates a genetic null phenotype. *Nat Biotechnol* **21**, 559–61.

17. Carmell, M. A., Zhang, L., Conklin, D. S., Hannon, G. J., and Rosenquist, T. A. (2003) Germline transmission of RNAi in mice. *Nat Struct Biol* **10**, 91–5.

18. Tiscornia, G., Tergaonkar, V., Galimi, F., Verma, I. M. (2004) CRE recombinase-inducible RNA interference mediated by lentiviral vectors. *Proc Natl Acad Sci U S A* **101**, 7347–51.

19. Szulc, J., and Aebischer, P. (2008) Conditional gene expression and knockdown using lentivirus vectors encoding shRNA. *Methods Mol Biol* **434**, 291–309.

20. Noseworthy, J. H., Lucchinetti, C., Rodriguez, M., and Weinshenker, B. G. (2000) Multiple sclerosis. *N Engl J Med* **343**, 938–52.

21. Compston, A., and Coles, A. (2002) Multiple sclerosis. *Lancet* **359**, 1221–31.

22. Debouverie, M., Pittion-Vouyovitch, S., Louis, S., and Guillemin, F. (2008) Natural history of multiple sclerosis in a population-based cohort. *Eur J Neurol* **15**, 916–21.

23. Rosati, G. (2001) The prevalence of multiple sclerosis in the world: an update. *Neurol Sci* **22**, 117–39.

24. Ascherio, A., and Munger, K. L. (2007) Environmental risk factors for multiple sclerosis. Part I: the role of infection. *Ann Neurol* **61**, 288–99.

25. Kornek, B., and Lassmann, H. (2003) Neuropathology of multiple sclerosis – new concepts. *Brain Res Bull* **61**, 321–26.

26. Svejgaard, A. (2008) The immunogenetics of multiple sclerosis. *Immunogenetics* **60**, 275–86.

27. International Multiple Sclerosis Genetics Consortium, Hafler, D. A., Compston, A., Sawcer, S., Lander, E. S., Daly, M. J., De Jager, P. L., de Bakker, P. I., Gabriel, S. B., Mirel, D. B., Ivinson, A. J., Pericak-Vance, M. A., Gregory, S. G., Rioux, J. D., McCauley, J. L., Haines, J. L., Barcellos, L. F., Cree, B., Oksenberg, J. R., and Hauser, S. L. (2007) Risk alleles for multiple sclerosis identified by a genomewide study. *N Engl J Med* **357**, 851–62.

28. Weber, F., Fontaine, B., Cournu-Rebeix, I., Kroner, A., Knop, M., Lutz, S., Müller-Sarnowski, F., Uhr, M., Bettecken, T., Kohli, M., Ripke, S., Ising, M., Rieckmann, P., Brassat, D., Semana, G., Babron, M. C., Mrejen, S., Gout, C., Lyon-Caen, O., Yaouanq, J., Edan, G., Clanet, M., Holsboer, F., Clerget-Darpoux, F., and Müller-Myhsok, B. (2008) IL2RA and IL7RA genes confer susceptibility for multiple sclerosis in two independent European populations. *Genes Immun* **9**, 259–63.

29. Anaya, J. M., Gómez, L., and Castiblanco, J. (2006) Is there a common genetic basis for autoimmune diseases? *Clin Dev Immunol* **13**, 185–95.

30. Spolski, R., Kashyap, M., Robinson, C., Yu, Z., and Leonard, W. J. (2008) IL-21 signaling is critical for the development of type I diabetes in the NOD mouse. *Proc Natl Acad Sci U S A* **105**, 14028–33.

31. Palacios, R., Aguirrezabal, I., Fernandez-Diez, B., Brieva, L., and Villoslada, P. (2005) Chromosome 5 and multiple sclerosis. *J Neuroimmunol* **167**, 1–3.

32. Richardson, J. H., Hofmann, W., Sodroski, J. G., and Marasco, W. A. (1998) Intrabody-mediated knockout of the high-affinity IL-2 receptor in primary human T cells using a bicistronic lentivirus vector. *Gene Ther* **5**, 635–44.

33. Gobin, S. J., Montagne, L., Van Zutphen, M., van der Valk, P., van den Elsen, P. J., and De Groot, C. J. (2001) Upregulation of transcription factors controlling MHC expression in multiple sclerosis lesions. *Glia* **36**, 68–77.

34. Frisullo, G., Mirabella, M., Angelucci, F., Caggiula, M., Morosetti, R., Sancricca, C., Patanella, A. K., Nociti, V., Iorio, R., Bianco, A., Tomassini, V., Pozzilli, C., Tonali, P. A., Matarese, G., and Batocchi, A. P. (2006) pSTAT1, pSTAT3, and T-bet expression in peripheral blood mononuclear cells from relapsing–remitting multiple sclerosis patients correlates with disease activity. *J Neurosci Res* **84**, 1027–36.

35. Cannella, B., and Raine, C. S. (2004) Multiple sclerosis: cytokine receptors on oligodendrocytes predict innate regulation. *Ann Neurol* **55**, 46–57.

36. David, M., Romero, G., Zhang, Z. Y., Dixon, J. E., and Larner, A. C. (1993) In vitro activation of the transcription factor ISGF3 by interferon alpha involves a membrane-associated tyrosine phosphatase and tyrosine kinase. *J Biol Chem* **268**, 6593–9.

37. Jiao, H., Berrada, K., Yang, W., Tabrizi, M., Platanias, L. C., and Yi, T. (1996) Direct association with and dephosphorylation of Jak2 kinase by the SH2-domain-containing protein tyrosine phosphatase SHP-1. *Mol Cell Biol* **16**, 6985–92.

38. Frank, C., Burkhardt, C., Imhof, D., Ringel, J., Zschörnig, O., Wieligmann, K., Zacharias, M., and Böhmer, F. D. (2004) Effective dephosphorylation of Src substrates by SHP-1. *J Biol Chem* **279**, 11375–83.

39. Massa, P. T., and Wu, C. (1996) The role of protein tyrosine phosphatase SHP-1 in the regulation of IFN-gamma signaling in neural cells. *J Immunol* **157**, 5139–44.

40. Massa, P. T., Saha, S., Wu, C., and Jarosinski, K. W. (2000) Expression and function of the protein tyrosine phosphatase SHP-1 in oligodendrocytes. *Glia* **29**, 376–85.

41. Massa, P. T., Wu, C., and Fecenko-Tacka, K. (2004) Dysmyelination and reduced myelin basic protein gene expression by oligodendrocytes of SHP-1-deficient mice. *J Neurosci Res* **77**, 15–25.

42. Christophil, G. P., Hudson, C. A., Gruber, R. C., Christophil, C. P., Mihai, C., Mejico, L. J., Jubelt, B., and Massa, P. T. (2008) SHP-1 deficiency and increased inflammatory gene expression in PBMCs of multiple sclerosis patients. *Lab Invest* **88**, 243–55.

43. Wrzesinski, S., Séguin, R., Liu, Y., Domville, S., Planelles, V., Massa, P., Barker, E., Antel, J., and Feuer, G. (2000) HTLV type 1 Tax transduction in microglial cells and astrocytes by lentiviral vectors. *AIDS Res Hum Retroviruses* **16**, 1771–6.

44. Fahn, S., Bressman, S. B., and Marsden, C. D. (1998) Classification of dystonia. *Adv Neurol* **78**, 1–10.

45. Ozelius, L. J., Hewett, J. W., Page, C. E., Bressman, S. B., Kramer, P. L., Shalish, C., de Leon, D., Brin, M. F., Raymond, D., Corey, D. P., Fahn, S., Risch, N. J., Buckler, A. J., Gusella, J. F., and Breakefield, X. O. (1997) The early-onset torsion dystonia gene (DYT1) encodes an ATP-binding protein. *Nat Genet* **17**, 40–8.

46. Hanson, P. I., and Whiteheart, S. W. (2005) AAA proteins: have engone, will work. *Nat Rev Mol Cell Biol* **6**, 519–52.

47. Hewett, J. W., Nery, F. C., Niland, B., Ge, P., Tan, P., Hadwiger, P., Tannous, B. A., Sah, D. W., and Breakefield, X. O. (2008) siRNA knockdown of mutant torsinA restores processing through secretory pathway in DYT1 dystonia cells. *Hum Mol Genet* **17**, 1436–45.

48. Kustedjo, K., Bracey, M. H. and Cravatt, B. F. (2000) Torsin A and its torsion dystonia-associated mutant forms are lumenal glycoproteins that exhibit distinct subcellular localizations. *J Biol Chem* **275**, 27933–9.

49. Gonzalez-Alegre, P., and Paulson, H. L. (2004) Aberrant cellular behavior of mutant torsinA implicates nuclear envelope dysfunction in DYT1 dystonia. *J Neurosci* **24**, 2593–601.

50. Goodchild, R. E. and Dauer, W. T. (2004) Mislocalization to the nuclear envelope: an effect of the dystonia-causing torsinA mutation. *Proc Natl Acad Sci U S A* **101**, 847–52.

51. Naismith, T. V., Heuser, J. E., Breakefield, X. O., and Hanson, P. I. (2004) TorsinA in the nuclear envelope. *Proc Natl Acad Sci U S A* **101**, 7612–7.

52. Bragg, D. C., Camp, S. M., Kaufman, C. A., Wilbur, J. D., Boston, H., Schuback, D. E., Hanson, P. I., Sena-Esteves, M., and Breakefield, X. O. (2004) Perinuclear biogenesis of mutant torsin-A inclusions in cultured cells infected with tetracycline-regulated herpes simplex virus type 1 amplicon vectors. *Neuroscience* **125**, 651–6.

53. Gonzalez-Alegre, P., Miller, V. M., Davidson, B. L., and Paulson, H. L. (2003) Toward therapy for DYT1 dystonia: allele-specific silencing of mutant TorsinA. *Ann Neurol* **53**, 781–7.

54. Hewett, J., Gonzalez-Agosti, C., Slater, D., Ziefer, P., Li, S., Bergeron, D., Jacoby, D. J., Ozelius, L. J., Ramesh, V., and Breakefield, X. O. (2000) Mutant torsinA, responsible for early-onset torsion dystonia, forms membrane inclusions in cultured neural cells. *Hum Mol Genet* **9**, 1403–13.

55. Gonzalez-Alegre, P., Bode, N., Davidson, B. L., and Paulson, H. L. (2005) Silencing primary dystonia: lentiviral-mediated RNA interference therapy for dyt1 dystonia. *J Neurosci* **25**, 10502–9.

56. Kock, N., Allchorne, A. J., Sena-Esteves, M., Woolf, C. J., and Breakefield, X. O. (2006) RNAi blocks DYT1 mutant torsinA inclusions in neurons. *Neurosci Lett* **395**, 201–5.

57. Nollen, E. A., Garcia, S. M., van Haaften, G., Kim, S., Chavez, A., Morimoto, R. I., and Plasterk, R. H. (2004) Genome-wide RNA interference screen identifies previously undescribed regulators of polyglutamine aggregation. *Proc Natl Acad Sci U S A* **101**, 6403–8.

58. Lieberman, A. P., and Fischbeck, K. H. (2000) Triplet repeat expansion in neuromuscular disease. *Muscle Nerve* **23**, 843–50.

59. Caplen, N. J., Taylor, J. P., Statha, V. S., Tanaka, F., Fire, A., and Morgan, R. A. (2002) Rescue of polyglutamine-mediated cytotoxicity by double-stranded RNA-mediated RNA interference. *Hum Mol Genet* **11**, 175–84.

60. Xia, H., Mao, Q., Eliason, S. L., Harper, S. Q., Martins, I. H., Orr, H. T., Paulson, H. L., Yang, L., Katin, R. M., and Davidson, B. L. (2004) RNAi suppresses polyglutamine-induced neurodegeneration in a model of spinocerebellar ataxia. *Nat Med* **10**, 816–20.

61. Alves, S., Régulier, E., Nascimento-Ferreira, I., Hassig, R., Dufour, N., Koeppen, A., Carvalho, A. L., Simões, S., de Lima, M. C., Brouillet, E., Gould, V. C., Déglon, N., and de Almeida, L. P. (2008) Striatal and nigral pathology in a lentiviral rat model of Machado–Joseph disease. *Hum Mol Genet* **17**, 2071–83.

62. Lavedan, C. (1998) The synuclein family. *Genome Res* **8**, 871–80.

63. Polymeropoulos, M., Lavedan, C., Leroy, E., Ide, S. E., Dehejia, A., Dutra, A., Pike, B., Root, H., Rubenstein, J., Boyer, R., Stenroos, E. S., Chandrasekharappa, S., Athanassiadou, A., Papapetropoulos, T., Johnson, W. G., Lazzarini, A. M., Duvoisin, R. C., Di Iorio, G., Golbe, L. I., and Nussbaum, R. L. (1997) Mutation in the alpha-synuclein gene identified in families with Parkinson's disease. *Science* **276**, 2045–7.

64. LoBianco, C., Ridet, J. L., Schneider, B. L., Deglon, N., and Aebischer, P. (2002) alpha-Synucleinopathy and selective dopaminergic neuron loss in a rat lentiviral-based model of Parkinson's disease. *Proc Nat Acad Sc U S A* **99**, 10813–8.

65. Fountaine, T. M., and Wade-Martins, R. (2007) RNA interference-mediated knockdown of alpha-synuclein protects human dopaminergic neuroblastoma cells from MPP(+) toxicity and reduces dopamine transport. *J Neurosci Res* **85**, 351–63.

66. Sapru, M. K., Yates, J. W., Hogan, S., Jiang, L., Halter, J., and Bohn, M. C. (2006) Silencing of human alpha-synuclein in vitro and in rat brain using lentiviral-mediated RNAi. *Exp Neurol* **198**, 382–90.

67. Peng, X. M., Tehranian, R., Dietrich, P., Stefanis, L., Perez, R. G. (2005) Alpha-synuclein activation of protein phosphatase 2A reduces tyrosine hydroxylase phosphorylation in dopaminergic cells. *J Cell Sci* **118**, 3523–30.

68. Perez, R. G., Waymire, J. C., Lin, E., Liu, J. J., Guo, F., Zigmond, M. J. (2002) A role for alpha-synuclein in the regulation of dopamine biosynthesis. *J Neurosci* **22**, 3090–9.

69. Alerte, T. N. M., Akinfolarin, A. A., Friedrich, E. E., Mader, S. A., Hong, C. S., and Perez, R. G. (2008) alpha-Synuclein aggregation alters tyrosine hydroxylase phosphorylation and immunoreactivity: lessons from viral transduction of knockout mice. *Neurosci Lett* **435**, 24–9.

70. Kieburtz, K., McDonald, M., Shih, C., Feigin, A., Steinberg, K., Bordwell, K., Zimmerman, C., Srinidhi, J., Sotack, J., Gusella, J., et al. (1994) Trinucleotide repeat length and progression of illness in Huntington's disease. *J Med Genet* **31**, 872–4.

71. Gusella, J. F., Wexler, N. S., Conneally, P. M., Naylor, S. L., Anderson, M. A., Tanzi, R. E., Watkins, P. C., Ottina, K., Wallace, M. R., and Sakaguchi, A. Y. et al. (1983) A polymorphic DNA marker genetically linked to Huntington's disease. *Nature* **306**, 234–8.

72. Bates, G., Harper, P., and Jones, L. (2002) Huntington's Disease, 3rd Edition. Oxford: Oxford University Press.

73. Imarisio, S., Carmichael, J., Korolchuk, V., Chen, C. W., Saiki, S., Rose, C., Krishna, G., Davies, J. E., Ttofi, E., Underwood, B. R., and Rubinsztein, D. (2008) Huntington's disease: from pathology and genetics to potential therapies. *Biochem J* **412**, 191–209.

74. Beal, M. F., and Ferrante, R. J. (2004) Experimental therapeutics in transgenic mouse models of Huntington's disease. *Nat Rev Neurosci* **5**, 373–84.

75. Kirik, D., and Björklund, A. (2003) Modeling CNS neurodegeneration by overexpression of disease-causing proteins using viral vectors. *Trends Neurosci* **26**, 386–92.

76. Ramaswamy, S., McBride, J. L., and Kordower, J. H. (2007) Animal models of Huntington's disease. *ILAR J* **48**, 356–73.

77. Harper, S. Q., Staber, P. D., He, X., Eliason, S. L., Martins, I. H., Mao, Q., Yang, L., Kotin, R. M., Paulson, H. L., and Davidson, B. L. (2005) RNA interference improves motor and neuropathological abnormalities in a Huntington's disease mouse model. Proc Natl Acad Sci U S A **102**, 5820–5.

78. McBride, J. L., Boudreau, R. L., Harper, S. Q., Staber, P. D., Monteys, A. M., Martins, I., Gilmore, B. L., Burstein, H., Peluso, R. W., Polisky, B., Carter, B. J., and Davidson, B. L. (2008) Artificial miRNAs mitigate shRNA-mediated toxicity in the brain: implications for the therapeutic development of RNAi. *Proc Natl Acad Sci U S A* **105**, 5868–73.

79. Zala, D., Bensadoun, J. C., Pereira de Almeida, L., Leavitt, B. R., Gutekunst, C. A., Aebischer, P., Hayden, M. R., and Déglon, N. (2004) Long-term lentiviral-mediated expression of ciliary neurotrophic factor in the striatum of Huntington's disease transgenic mice. *Exp Neurol* **185**, 26–35.

80. Zala, D., Benchoua, A., Brouillet, E., Perrin, V., Gaillard, M. C., Zurn, A. D., Aebischer, P., and Déglon, N. (2005) Progressive and selective striatal degeneration in primary neuronal cultures using lentiviral vector coding for a mutant huntingtin fragment. *Neurobiol Dis* 20, 785–98.

81. Runne, H., Regulier, E., Kuhn, A., Zala, D., Gokce, O., Perrin, V., Sick, B., Aebischer, P., Deglon, N., and Luthi-Carter, R. (2008) Dysregulation of gene expression in primary neuron models of huntington's disease shows that polyglutamine-related effects on the striatal transcriptome may not be dependent on brain circuitry. *J Neurosci* 28, 9723–31.

82. deAlmeida, L. P., Ross, C. A., Zala, D., Aebischer, P., and Déglon, N. (2002) Lentiviral-mediated delivery of mutant huntingtin in the striatum of rats induces a selective neuropathology modulated by polyglutamine repeat size, huntingtin expression levels, and protein length. *J Neurosci* 22, 3473–83.

83. Régulier, E., Trottier, Y., Perrin, V., Aebischer, P., and Déglon, N. (2003) Early and reversible neuropathology induced by tetracycline-regulated lentiviral overexpression of mutant huntingtin in rat striatum. *Hum Mol Genet* 12, 2827–36.

84. Perrin, V., Régulier, E., Abbas-Terki, T., Hassig, R., Brouillet, E., Aebischer, P., Luthi-Carter, R., and Déglon, N. (2007) Neuroprotection by Hsp104 and Hsp27 in lentiviral-based rat models of Huntington's disease. *Mol Ther* 15, 903–11.

85. Popovic, N., Maingay, M., Kirik, D., and Brundin, P. (2005) Lentiviral gene delivery of GDNF into the striatum of R6/2 Huntington mice fails to attenuate behavioral and neuropathological changes. *Exp Neurol* 193, 65–74.

86. Benchoua, A., Trioulier, Y., Zala, D., Gaillard, M. C., Lefort, N., Dufour, N., Saudou, F., Elalouf, J. M., Hirsch, E., Hantraye, P., Déglon, N., and Brouillet, E. (2006) Involvement of mitochondrial complex II defects in neuronal death produced by N-terminus fragment of mutated huntingtin. Mol Biol Cell 17, 1652–63.

87. Benchoua, A., Trioulier, Y., Diguet, E., Malgorn, C., Gaillard, M. C., Dufour, N., Elalouf, J. M., Krajewski, S., Hantraye, P., Deglon, N., and Brouillet, E. (2008) Dopamine determines the vulnerability of striatal neurons to the N-terminal fragment of mutant huntingtin through the regulation of mitochondrial complex II. *Hum Mol Genet* 17, 1446–56.

88. Charvin, D., Roze, E., Perrin, V., Deyts, C., Betuing, S., Pagès, C., Régulier, E., Luthi-Carter, R., Brouillet, E., Déglon, N., and Caboche, J. (2008) Haloperidol protects striatal neurons from dysfunction induced by mutated huntingtin in vivo. *Neurobiol Dis* 29, 22–9.

89. Cui, L., Jeong, H., Borovecki, F., Parkhurst, C. N., Tanese, N., and Krainc, D. (2007) Transcriptional repression of PGC-1alpha by mutant huntingtin leads to mitochondrial dysfunction and neurodegeneration. *Cell* 127, 59–69.

90. Fukui, H., and Moraes, C. T. (2007) Extended polyglutamine repeats trigger a feedback loop involving the mitochondrial complex III, the proteasome and huntingtin aggregates. *Hum Mol Genet* 16, 783–97.

91. Dass, B., and Kordower, J. H. (2007) Gene therapy approaches for the treatment of Parkinson's disease. *Handb Clin Neurol* 84, 291–304.

92. Déglon, N., Tseng, J. L., Bensadoun, J. C., Zurn, A. D., Arsenijevic, Y., deAlmeida, L., Zufferey, R., Trono, D., and Aebischer, P. (2000) Self-inactivating lentiviral vectors with enhanced transgene expression as potential gene transfer system in Parkinson's disease. *Hum Gene Ther* 11, 179–90.

93. Dowd, E., Monville, C., Torres, E. M., Wong, L. F., Azzouz, M., Mazarakis, N. D., and Dunnett, S. B. (2005) Lentivector-mediated delivery of GDNF protects complex motor functions relevant to human Parkinsonism in a rat lesion model. *Eur J Neurosci* 22, 2587–95.

94. Brizard, M., Carcenac, C., Bemelmans, A. P., Feuerstein, C., Mallet, J., and Savasta, M. (2006) Functional reinnervation from remaining DA terminals induced by GDNF lentivirus in a rat model of early Parkinson's disease. *Neurobiol Dis* 21, 90–101.

95. Kitada, T., Asakawa, S., Hattori, N., Matsumine, H., Yamamura, Y., Minoshima, S., Sokochi, M., Mizuno, Y., and Shimizu, N. (1998) Mutations in the parkin gene cause autosomal recessive juvenile parkinsonism. *Nature* 392, 605–8.

96. Imai, Y., Soda, M., and Takahashi, R. (2000) Parkin suppresses unfolded protein stress induced cell death through its E3 ubiquitin-protein ligase activity. *J Biol Chem* 275, 35661–4.

97. Doss-Pepe, E. W., Chen, L., and Madura, K. (2005) Alpha-synuclein and parkin contribute to the assembly of ubiquitin lysine 63-linked multiubiquitin chains. *J Biol Chem* 280, 16619–24.

98. Lim, K. L., Chew, K. C., Tan, J. M., Wang, C., Chung, K. K., Zhang, Y., Tanaka, Y., Smith, W., Engelender, S., Ross, C. A., Dawson, V. L., and Dawson, T. M. (2005) Parkin mediates nonclassical, proteasomal-independent ubiquitination of synphilin-1:

implications for Lewy body formation. *J Neurosci* **25**, 2002–9.

99. Ulusoy, A., and Kirik, D. (2008) Can overexpression of parkin provide a novel strategy for neuroprotection in Parkinson's disease? *Exp Neurol* **212**, 258–60.

100. LoBianco, C., Schneider, B. L., Bauer, M., Sajadi, A., Brice, A., Iwatsubo, T., and Aebischer, P. (2004) Lentiviral vector delivery of parkin prevents dopaminergic degeneration in an alpha-synuclein rat model of Parkinson's disease. *Proc Nat Acad Sci U S A* **101**, 17510–5.

101. Ridet, J. L., Bensadoun, J. C., Déglon, N., Aebischer, P., and Zurn, A. D. (2006) Lentivirus-mediated expression of glutathione peroxidase: Neuroprotection in murine models of Parkinson's disease. *Neurobiol Dis* **21**, 29–34.

102. Vergo, S., Johansen, J. L., Leist, M., and Lotharius, J. (2007) Vesicular monoamine transporter 2 regulates the sensitivity of rat dopaminergic neurons to disturbed cytosolic dopamine levels. *Brain Res* **1185**, 18–32.

103. Deierborg, T., Soulet, D., Roybon, L., Hall, V., and Brundin, P. (2008) Emerging restorative treatments for Parkinson's disease. *Prog Neurobiol* **85**, 407–32.

104. Chao, C. C., and Lee, E. H. Y. (1999) Neuroprotective mechanism of glial cell line-derived neurotrophic factor on dopamine neurons: role of antioxidation. *Neuropharmacology* **38**, 913–6.

105. Kordower, J. H., Emborg, M. E., Bloch, J., Ma, S. Y., Chu, Y., Leventhal, L., McBride, J., Chen, E. Y., Palfi, S., Roitberg, B. Z., Brown, W. D., Holden, J. E., Pyzalski, R., Taylor, M. D., Carvey, P., Ling, Z., Trono, D., Hantraye, P., Deglon, N., and Aebischer, P. (2000) Neurodegeneration prevented by lentiviral vector delivery of GDNF in primate models of Parkinson's disease. *Science* **290**, 767–73.

106. Jakobsson, J., and Lundberg, C. (2006) Lentiviral vectors for use in the central nervous system. *Mol Ther* **13**, 484–93.

107. Bilang-Bleuel, A., Revah, F., Colin, P., Locquet, I., Robert, J. J., Mallet, J., and Horellou, P. (1997) Intrastriatal injection of an adenoviral vector expressing glial-cell-line-derived neurotrophic factor prevents dopaminergic neuron degeneration and behavioral impairment in a rat model of Parkinson disease. *Proc Natl Acad Sci U S A* **94**, 8818–23.

108. Rosenblad, C., Gronborg, M., Hansen, C., Blom, N., Meyer, M., Johansen, J., Dago, L., Kirik, D., Patel, U. A., Lundberg, C., Trono, D., Bjorklund, A., and Johansen, T.

E. (2000) *In vivo* protection of nigral dopamine neurons by lentiviral gene transfer of the novel GDNF-family member neublastin/artemin3. *Mol Cell Neurosci* **15**, 199–214.

109. Georgievska, B., Kirik, D., Rosenblad, C., Lundberg, C., and Bjorklund, A. (2002) Neuroprotection in the rat Parkinson model by intrastriatal GDNF gene transfer using a lentiviral vector. *NeuroReport* **13**, 75–82.

110. Azzouz, M., Ralph, S., Wong, L. F., Day, D., Askham, Z., Barber, R. D., Mitrophanous, K. A., Kingsman, S. M., and Mazarakis, N. D. (2004) Neuroprotection in a rat Parkinson model by GDNF gene therapy using EIAV vector. *Neuroreport* **15**, 985–90.

111. Palfi, S., Leventhal, L., Chu, Y., Ma, S. Y., Emborg, M., Bakay, R., Deglon, N., Hantraye, P., Aebischer, P., and Kordower, J. H. (2002) Lentivirally delivered glial cell line-derived neurotrophic factor increases the number of striatal dopaminergic neurons in primate models of nigrostriatal degeneration. *J Neurosci* **22**, 4942–54.

112. Winkler, C., Sauer, H., Lee, C. S., and Bjorklund, A. (1996) Short-term GDNF treatment provides long-term rescue of lesioned nigral dopaminergic neurons in a rat model of Parkinson's disease. *J Neurosci* **16**, 7206–15.

113. Bjorklund, A., Kirik, D., Rosenblad, C., Georgievska, B., Lundberg, C., and Mandel, R. J. (2000) Towards a neuroprotective gene therapy for Parkinson's disease: use of adenovirus, AAV and lentivirus vectors for gene transfer of GDNF to the nigrostriatal system in the rat Parkinson model. *Brain Res* **886**, 82–98.

114. Choi-Lundberg, D. L., Lin, Q., Schallert, T., Crippens. D., Davidson, B. L., Chang, Y. N., Chiang, Y. L., Qian, J., Bardwaj, L., and Bohn, M. C.(1998) Behavioral and cellular protection of rat dopaminergic neurons by an adenoviral vector encoding glial cell line-derived neurotrophic factor. *Exp Neurol* **154**, 261–75.

115. LoBianco, C., Déglon, N., Pralong, W., and Aebischer, P. (2004) Lentiviral nigral delivery of GDNF does not prevent neurodegeneration in a genetic rat model of Parkinson's disease. *Neurobiol Dis* **17**, 283–9.

116. Capowski, E. E., Schneider, N. L., Ebert, A. D., Seehus, C. R., Szulc, J., Zufferey, R., Aebischer, P., and Svendsen, C. N. (2007) Lentiviral vector-mediated genetic modification of human neural progenitor cells for ex vivo gene therapy. *J Neurosci Methods* **163**, 338–49.

117. Ebert, A. D., Beres, A. J., Barber, A. E., and Svendsen, C. N. (2008) Human neural progenitor cells over-expressing IGF-1 protect dopamine neurons and restore function in a

rat model of Parkinson's disease. *Exp Neurol* **209**, 213–23.

118. Caplen, A. (2000) Gene therapy for neuro-degeneration. *Trends Mol Med* **7**, 51–5.

119. Takasugi, N., Takahashi, Y., Morohashi, Y., Tomita, T., and Iwatsubo, T. (2002) The mechanism of gamma-secretase activities through high molecular weight complex formation of presenilins is conserved in Drosophila melanogaster and mammals. *J Biol Chem* **277**, 50198–205.

120. Kao, S. C., Krichevsky, A. M., Kosik, K. S., and Tsai, L. H. (2004) BACE1 suppression by RNA interference in primary cortical neurons. *J Biol Chem* **279**, 1942–9.

121. Vassar, R. (2004) BACE 1: the beta-secretase enzyme in Alzheimer's disease. *J Mol Neurosci* **23**, 105–14.

122. Burton, A. (2005) RNA inhibition of beta-secretase reverts AD in mice. *Lancet Neurol* **4**, 698.

123. Sierant, M., Kubiak, K., Kazmierczak-Baranska, J., Paduszynska, A., Kuwabara, T., Warashina, M., Nacmias, B., Sorbi, S., and Nawrot, B. (2008) RNA interference in silencing of genes of Alzheimer's disease in cellular and rat brain models. *Nucleic Acids Symp Ser (Oxf)* **52**, 41–2.

124. Singer, O., Marr, R. A., Rockenstein, E., Crews, L., Coufal, N. G., Gage, F. H., Verma, I. M., and Masliah, E. (2005) Targeting BACE1 with siRNAs ameliorates Alzheimer disease neuropathology in a transgenic model. *Nat Neurosci* **8**, 1343–9.

125. Sun, X., He, G., and Song, W. (2006) BACE2, as a novel APP theta-secretase, is not responsible for the pathogenesis of Alzheimer's disease in Down syndrome. *FASEB J* **20**, 1369–76.

126. El-Amouri, S. S., Zhu, H., Yu, J., Marr, R., Verma, I. M., and Kindy, M. S. (2008) Neprilysin: an enzyme candidate to slow the progression of Alzheimer's disease. *Am J Pathol* **172**, 1342–54.

127. Mueller-Steiner, S., Zhou, Y., Arai, H., Roberson, E. D., Sun, B., Chen, J., Wang, X., Yu, G., Esposito, L., Mucke, L., and Gan, L. (2006) Antiamyloidogenic and neuroprotective functions of cathepsin B: implications for Alzheimer's disease. *Neuron* **51**, 703–14.

128. Cole, G., and Frautschy, S. A. (2006) Cat and mouse. *Neuron* **51**(6), 671–2.

129. Dodart, J. C., Marr, R. A., Koistinaho, M., Gregersen, B. M., Malkani, S., Verma, I. M., and Paul, S. M. (2005) Gene delivery of human apolipoprotein E alters brain Abeta burden in a mouse model of Alzheimer's disease. *Proc Natl Acad Sci U S A* **102**, 1211–6.

130. Kim, D., Nguyen, M. D., Dobbin, M. M., Fischer, A., Sananbenesi, F., Rodgers, J. T., Delalle, I., Baur, J. A., Sui, G., Armour, S. M., Puigserver, P., Sinclair, D. A., and Tsai, L. H. (2007) SIRT1 deacetylase protects against neurodegeneration in models for Alzheimer's disease and amyotrophic lateral sclerosis. *EMBO J* **26**, 369–79.

131. Chen, J., Zhou, Y., Mueller-Steiner, S., Chen, L. F., Kwon, H., Yi, S., Mucke, L., and Gan, L. (2005) SIRT1 protects against microglia-dependent amyloid-beta toxicity through inhibiting NF-kappaB signaling. *J Biol Chem* **280**, 40364–74.

132. Richard, K. L., Filali, M., Préfontaine, P., and Rivest, S. (2008) Toll-like receptor 2 acts as a natural innate immune receptor to clear amyloid beta 1-42 and delay the cognitive decline in a mouse model of Alzheimer's disease. *J Neurosci* **28**, 5784–93.

133. Nestler, E. (2000) Genes and addiction. *Nat Genet* **26**, 277–81.

134. Bahi, A., Boyer, F., Bussard, G., and Dreyer, J. L. (2005) Silencing dopamine D3-receptor in the nucleus accumbens shell *in vivo* induces behavioral changes in chronic cocaine delivery. *Eur J Neurosci* **21**, 3415–26.

135. Bahi, A., Boyer, F., and Dreyer, J. L. (2008) Cocaine-induced behavioral and reward upon lentivirus-mediated expression changes of BDNF and TrkB in the nucleus accumbens. *Psychopharmacology* **200**, 129–39.

136. Wersinger, C., Prou, D., Vernier, P., and Sidhu, A. (2003) Modulation of dopamine transporter function by alpha-synuclein is altered by impairment of cell adhesion and by induction of oxidative stress. *FASEB J* **17**, 2151–3.

137. Wersinger, C., and Sidhu, A. (2005) Disruption of the interaction of alpha-synuclein with microtubules enhances cell surface recruitment of the dopamine transporter. *Biochemistry* **44**, 13612–24.

138. Boyer, F., and Dreyer, J. L. (2007) Alpha-synuclein in the nucleus accumbens induces changes in cocaine behavior in rats. *Eur J Neurosci* **26**, 2764–76.

139. Boyer, F., and Dreyer, J. L. (2008) The role of gamma-synuclein in cocaine-induced behavior in rats. *Eur J Neurosci* **27**, 2938–51.

140. Robinson, T. E., and Berridge, K. C. (1993) The neural basis of drug craving: an incentive-sensitization theory of addiction. *Brain Res Brain Res Rev* **18**, 247–291.

141. Yue, Y., Chen, Z. Y., Gale, N. W. Blair-Flynn, J., Hu, T. J., Yue, X., Cooper, M., Crockett, D. P., Yancopoulos, G. D., Tessarollo, L., and Zhou, R. (2002) Mistargeting hippocampal axons

by expression of a truncated Eph receptor. *Proc Natl Acad Sci U S A* **99**, 10777–82.

142. Bahi, A., and Dreyer, J. L. (2005) Cocaine-induced expression changes of axon guidance molecules in the adult rat brain. *Mol Cell Neurosci* **28**, 275–91.

143. Halladay, A. K., Yue, Y., Michna, L., Widmer, D. A., Wagner, G. C., Zhou, R. (2000) Regulation of EphB1 expression by dopamine signaling. *Brain Res Mol Brain Res* **85**, 171–8.

144. Brenz-Verca, M. S., Widmer, D. A. J., Wagner, G. C., and Dreyer, J. L. (2001) Cocaine-induced expression of the tetratspanin CD81 and its relation to hypothalamic function. *Mol Cell Neurosci* **17**, 303–16.

145. Michna, L., Brenz-Verca, M. S., Widmer, D. A. J., Chen, S., Lee, J., Rogove, J., Zhou, R., Tsitsikov, E., Miescher, G. C., Dreyer, J. L., and Wagner, G. C. (2001) Altered sensitivity of CD81-deficient mice to neurobehavioral effects of cocaine. *Mol Brain Res* **90**, 68–74.

146. Bahi, A., Boyer, F., Kafri, T., and Dreyer, J. L. (2004) CD81-induced behavioural changes during chronic cocaine administration: *in vivo* gene delivery with regulatable lentivirus. *Eur J Neurosci* **19**, 1621–33.

147. Bahi, A., Boyer, F., Kolira, M., and Dreyer, J. L. (2005) *In vivo* gene silencing of CD81 by lentiviral expression of siRNAs suppresses cocaine-induced behavior. *J Neurochem* **92**, 1243–55.

148. Bahi, A., Boyer, F., Gumy, C., Kafri, T., and Dreyer, J. L. (2004) *In vivo* gene delivery of urokinase-type plasminogen activator with regulatable lentivirus induces behavioural changes in chronic cocaine administration. *Eur J Neurosci* **20**, 3473–88.

149. Bahi, A., Boyer, F., and Dreyer, J. L. (2006) Silencing urokinase in the ventral tegmental area *in vivo* induces changes in cocaine-induced hyperlocomotion. *Eur J Neurosci* **98**, 1619–31.

150. Bahi, A., Kusnecov, A., and Dreyer, J. L. (2008) Effects of Urokinase-type plasminogen activator in the acquisition, expression and reinstatement of cocaine-induced conditioned place preference. *Behav Brain Res* **191**, 17–25.

151. Bahi, A., and Dreyer, J. L. (2008) Overexpression of plasminogen activators in the nucleus accumbens enhances cocaine, amphetamine and morphine-induced reward and behavioral sensitization. *Genes Brain Behav* **7**, 244–56.

152. Bahi, A., Kusnecov, A., and Dreyer, J. L. (2008) The role of tissue-type plasminogen activator system in amphetamine-induced conditioned place preference extinction and reinstatement. *Neuropschopharmacology* **33**, 2726–34.

153. Yan, Y., Yamada, K., Mizoguchi, H., Noda, Y., Nagai, T., Nitta, A., and Nabeshima, T. (2007) Reinforcing effects of morphine are reduced in tissue plasminogen activator-knockout mice. *Neuroscience* **146**, 50–9.

154. Dulcan, M. (1997) Practice parameters for the assessment and treatment of children, adolescents, and adults with attention-deficit/hyperactivity disorder. American Academy of Child and Adolescent Psychiatry. *J Am Acad Child Adolesc Psychiatry* **36**(Suppl 10), 85S–121S.

155. Doyle, A. E. (2006) Executive functions in attention-deficit/hyperactivity disorder. *J Clin Psychiatry* **67**, 21–6.

156. Castellanos, F. X., Sonuga-Barke, E. J., Milham, M. P., and Tannock, R. (2006) Characterizing cognition in ADHD: beyond executive dysfunction. *Trends Cogn Sci* **10**, 117–23.

157. Willcutt, E. G., Pennington, B. F., Olson, R. K., Chhabildas, N., and Hulslander, J. (2005) Neuropsychological analyses of comorbidity between reading disability and attention deficit hyperactivity disorder: in search of the common deficit. *Dev Neuropsychol* **27**, 35–78.

158. Sonuga-Barke, E. J. (2005) Editorial. *J Child Psychol Psychiatry* **46**, 225–6.

159. Sagvolden, T., and Sergeant, J. A. (1998) Attention deficit/hyperactivity disorder – from brain dysfunctions to behaviour. *Behav Brain Res* **94**, 1–10.

160. Oades, R. D. (1998) Frontal, temporal and lateralized brain function in children with attention-deficit hyperactivity disorder: a psychophysiological and neuropsychological viewpoint on development. *Behav Brain Res* **94**, 83–95.

161. Black, D. W., and Moyer, T. (1998) Clinical features and psychiatric comorbidity of subjects with pathological gambling behavior. *Psychiatr Serv* **49**, 1434–9.

162. Comings, D. E. (2001) Clinical and molecular genetics of ADHD and Tourette syndrome. Two related polygenic disorders. *Ann N Y Acad Sci* **931**, 50–83.

163. Swanson, J. M., Flodman, P., Kennedy, J., Spence, M. A., Moyzis, R., Schuck, S., Murias, M., Moriarity, J., Barr, C., Smith, M., and Posner, M. (2000) Dopamine genes and ADHD. *Neurosci Biobehav Rev* **24**, 21–5.

164. Smith, K. M., Daly, M., Fischer, M., Yiannoutsos, C. T., Bauer, L., Barkley, R.,

Navia, B. A. (2003) Association of the dopamine beta hydroxylase gene with attention deficit hyperactivity disorder: genetic analysis of the Milwaukee longitudinal study. *Am J Med Genet B Neuropsychiatr Genet* **119B**, 77–85.

165. Roman, T., Rohde, L. A., and Hutz, M. H. (2004) Polymorphisms of the dopamine transporter gene: influence on response to methylphenidate in attention deficit-hyperactivity disorder. *Am J Pharmacogenomics* **4**, 83–92.

166. Oades, R. D. (2008) Dopamine-serotonin interactions in attention-deficit hyperactivity disorder (ADHD). *Prog Brain Res* **172**, 543–65.

167. Adriani, W., Boyer, F., Gioiosa, L., Macrì, S., Dreyer, J. L., and Laviola, G. (2008) Increased impulsive behavior and gambling temptation following lentivirus-mediated DAT overexpression in rats' nucleus accumbens. *Neuroscience* **159**, 47–58.

168. Laviola, G., Macrì, S., Morley-Fletcher, S., and Adriani, W. (2003) Risk-taking behavior in adolescent mice: psychobiological determinants and early epigenetic influence. *Neurosci Biobehav Rev* **27**, 19–31.

169. Adriani, W., and Laviola, G. (2006) Delay aversion but preference for large and rare rewards in two choice tasks: implications for the measurement of self-control parameters. *BMC Neurosci* **7**, 52.

170. Jin, J., Bao, X., Wang, H., Pan, H., Zhang, Y., and Wu, X. (2008) RNAi-induced downregulation of Mecp2 expression in the rat brain. *Int J Dev Neurosci* **26**, 457–65.

171. Nelson, E. D., Kavalali, E. T., and Monteggia, L. M. (2006) MeCP2-dependent transcriptional repression regulates excitatory neurotransmission. *Curr Biol* **16**, 710–6.

172. Raoul, C., Abbas-Terki, T., Bensadoun, J. C., Guillot, S., Haase, G., Szulc, J., Henderson, C. E., and Aebischer, P. (2005) Lentiviral-mediated silencing of SOD1 through RNA interference retards disease onset and progression in a mouse model of ALS. *Nat Med* **11**, 423–8.

173. Ralph, G. S., Radcliffe, P. A., Day, D. M., Carthy, J. M., Leroux, M. A., Lee, D. C. P., Wong, L. F., Bilsland, L. G., Greensmith, L., Kingsman, S. M., Mitrophanous, K. A., Mazarakis, N. D., and Azzouz. M. (2005) Silencing mutant SOD1 using RNAi protects against neurodegeneration and extends survival in an ALS model. *Nat Med* **11**, 429–33.

174. Pfeifer, A., Eigenbrod, S., Al-Khadra, S., Hofmann, A., Mitteregger, G., Moser, M., Bertsch, U., and Kretzschma, H. (2006) Lentivector-mediated RNAi efficiently suppresses prion protein and prolongs survival of scrapie-infected mice. *J Clin Invest* **116**, 3204–10.

175. Zhao, P., Wang, C., Fu, Z., You, Y., Cheng, Y., Lu, X., Lu, A., Liu, N., Pu, P., Kang, C., Salford, L. G., and Fan, X. (2008) Lentiviral vector mediated siRNA knockdown of hTERT results in diminished capacity in invasiveness and *in vivo* growth of human glioma cells in a telomere length-independent manner. *Int J Oncol* **31**, 361–8.

176. Hendriks, W. T. J., Ruitenberg, M. J., Blits, B., Boer, G. J., and Verhaagen, J. (2004) Viral vector-mediated gene transfer of neurotrophins to promote regeneration of the injured spinal cord. *Prog Brain Res* **146**, 451–76.

177. Wu, D., Zhang, Y., Bo, X., Huang, W., Xiao, F., Zhang, X., Miao, T., Magoulas, C., and Subang, M. C. (2007) Actions of neuropoietic cytokines and cyclic AMP in regenerative conditioning of rat primary sensory neuron. *Exp Neurol* **204**, 66–76.

178. Dittgen, T., Nimmerjahn, A., Komai, S., Licznerski, P., Waters, J., Margrie, T. W., Helmchen, F., Denk, W., Brecht, M., and Osten, P. (2004) Lentivirus-based genetic manipulations of cortical neurons and their optical and electrophysiological monitoring *in vivo*. *Proc Nat Acad Sci U S A* **101**, 18206–11.

179. Kameda, H., Furuta, T., Matsuda, W., Ohira, K., Nakamura, K., Hioki, H., Kaneko, T. (2008) Targeting green fluorescent protein to dendritic membrane in central neurons. *Neurosci Res* **61**, 79–91.

180. Santamaria, J., Khalfallah, O., Sauty, C., Brunet, I., Sibieude, M., Mallet, J., Berrard, S., and Lecomte, M. J. (2009) Silencing of choline acetyltransferase expression by lentivirus-mediated RNA interference in cultured cells and in the adult rodent brain. *J Neurosci Res* **87**(2), 532–44.

181. Crittenden, J. R., Heidersbach, A., and McManus, M. T. (2007) Lentiviral strategies for RNAi knockdown of neuronal genes. *Curr Protoc Neurosci* **5**, 5–26.

182. Porras, G., and Bezard, E. (2008) Preclinical development of gene therapy for Parkinson's disease. *Exp Neurol* **209**, 72–81.

Part II

Advances in Lentiviral Vector Technology

Chapter 2

New Protocol for Lentiviral Vector Mass Production

María Mercedes Segura, Alain Garnier, Yves Durocher, Sven Ansorge, and Amine Kamen

Abstract

Multiplasmid transient transfection is the most widely used technique for the generation of lentiviral vectors. However, traditional transient transfection protocols using 293 T adherent cells and calcium phosphate/DNA co-precipitation followed by ultracentrifugation are tedious, time-consuming, and difficult to scale up. This chapter describes a streamlined protocol for the fast mass production of lentiviral vectors and their purification by affinity chromatography. Lentiviral particles are generated by transient transfection of suspension growing HEK 293 cells in serum-free medium using polyethylenimine (PEI) as transfection reagent. Lentiviral vector production is carried out in Erlenmeyer flasks agitated on orbital shakers requiring minimum supplementary laboratory equipment. Alternatively, the method can be easily scaled up to generate larger volumes of vector stocks in bioreactors. Heparin affinity chromatography allows for selective concentration and purification of lentiviral particles in a singlestep directly from vector supernatants. The method is suitable for the production and purification of different vector pseudotypes.

Key words: Lentiviral vector production, Lentiviral vector purification, Scaleable processes, Transient transfection, Cell suspension culture, Serum-free medium, Chromatography

1. Introduction

Lentiviral vectors play a key role as gene delivery vehicles in many fundamental and applied research applications since, unlike most currently existing gene transfer systems, they are able to provide long-term gene expression. Along with the growing interest in lentiviral vectors, comes the need to develop streamlined production and purification procedures for the generation of high-titer vector stocks. A variety of production systems have shown to efficiently generate transduction-competent lentiviral particles, including transient transfection of mammalian cells (1), the use of stable packaging cell lines (2), and more recently, baculovirus

Maurizio Federico (ed.), *Lentivirus Gene Engineering Protocols,* Methods in Molecular Biology, vol. 614,
DOI 10.1007/978-1-60761-533-0_2, © Humana Press, a part of Springer Science+Business Media, LLC 2010

technology (3). However, multiplasmid transient transfection continues to be the most commonly used technique for the generation of lentiviral stocks, since it constitutes a faster, simpler, and more versatile approach. Transient transfection avoids the time-consuming and tedious process of developing stable packaging cell lines or recombinant baculoviruses prior to lentiviral vector production. In addition, it allows for the use of cytotoxic/cytostatic transgenes and/or vector components, which is the case for many HIV-1-derived proteins (4–7) and for the commonly used vesicular stomatitis virus pseudotyping envelope glycoprotein (VSV-G) (8), that otherwise need to be put under tight regulatory control (9). Furthermore, transient transfection production systems permit testing various transgenes of interest and envelope proteins with alternative cell tropisms in a relatively short time (10).

Detailed protocols describing the production of lentiviral stocks by multiplasmid transient transfection of human embryonic kidney (HEK) 293 T cells using calcium phosphate/DNA co-precipitation are available (11–13). In these protocols, producer cells are grown as monolayers in 10 or 15 cm culture dishes in the presence of 10% fetal bovine serum (FBS). Following successful transfection, 10–15 mL of supernatant containing ~10^6–10^7 infective lentiviral particles per mL (IVP/mL) can be recovered at day 2 and 3 posttransfection. Subsequently, harvested supernatants are pooled, filtered through 0.45-μm membranes, and usually purified through two rounds of ultracentrifugation in order to improve vector potency and purity. A critical parameter for correct calcium phosphate/DNA co-precipitation is the pH. A strict control of the pH of the reagents used for transfection and the percentage of CO_2 in the incubators is required because slight variations in the pH can lead to the formation of a too fine or too coarse precipitate that will adversely affect transfection efficiency (12–14). Transfection efficiency may also be affected by the lot of FBS used for cell culture (13). Another problem frequently encountered using these protocols is the loose attachment of HEK 293 T producer cells to culture dishes (13–15). This greatly complicates the production process, as extreme care needs to be taken to prevent producer cell detachment during wash and medium replacement steps. In some protocols, this problem has been circumvented by precoating culture dishes with poly-L-lysine (12). It is also important to change the culture medium after transfection since calcium phosphate is toxic to the producer cells, which can further contribute to HEK 293 T cell detachment (14). In order to overcome the practical difficulties associated with these protocols, reduce the time and effort required for vector production and purification, and allow method scalability, we have developed and optimized a new protocol for lentiviral vector production (16).

This protocol describes how to produce lentiviral vectors by polyethylenimine (PEI)-mediated transient transfection of HEK 293 cells grown in suspension culture and serum-free conditions. Protocols for the fast titration and chromatography purification of lentiviral vector stocks are also provided. Transient transfection is achieved using PEI, a cationic polymer, that binds DNA resulting in the formation of a compact DNA–PEI complex (polyplex) that can efficiently transfect mammalian cells. The use of PEI has a number of advantages over calcium phosphate/DNA coprecipitation. It allows for efficient transfection of HEK 293 cells, both in adherent and suspension cultures while showing minimal cytotoxic effects. It is much easier to use and does not require a tight control of transfection conditions, which are difficult to achieve at large scale (17). PEI-mediated transfection is effective, both in the presence or absence of serum in the culture medium. Other efficient transfection reagents, such as cationic lipids (e.g., lipofectamine™), are too expensive particularly when considering large-scale vector production. In this protocol, we describe the production of lentiviral vectors by transient transfection using a third generation lentiviral vector system. This system, specifically designed to minimize the risk of generating replication-competent virus (RCV) (18), consists of four plasmids: two packaging plasmids coding for Gag/Pol and Rev sequences, a VSV-G plasmid, and a transfer vector plasmid coding for the green fluorescent protein (GFP) marker to facilitate determination of transfection efficiencies and viral titers. It is important to mention that the method has been successfully adapted to the production of lentiviral vectors using a different combination and number of plasmids (unpublished results).

One key aspect for the success of this protocol is the use of HEK 293 cells adapted to grow in suspension. Production of lentiviral vectors using adherent cell cultures can only be scaled up by increasing cell attachment surface. This is typically achieved in the laboratory by increasing the number of dishes used for vector production, which results in a tedious and time-consuming vector production process. Scale-up can also be accomplished by using alternative culture devices (e.g., roller bottles, multitray cell factories, or microcarriers in stirred tanks) that provide extended anchorage surface. However, these cell culture systems are complex, costly, and only provide a limited scalability. To overcome these problems, a suspension-adapted HEK 293 cell line is used in our laboratory for lentiviral vector production. Suspension-adapted cells can easily be grown at high cell densities in stirred tanks, which can be scaled up to several thousands of liters. The use of suspension growing cells also avoids potential problems associated with loose attachment of HEK 293 cells to culture dishes and the use of trypsin, which further simplifies the production technique. Lentiviral vector production is carried out in Erlenmeyer flasks agitated on orbital shakers requiring minimum supplementary laboratory equipment (CO_2 and humidity-resistant orbital shaker). The production scale described in this protocol is small (125 mL shake flasks containing 20 mL of cell culture). The procedure can be easily scaled up in the laboratory to 650 mL cell culture in 2-L shake flasks. Larger volumes of vector supernatant have been successfully produced in bioreactors (16). The method can also be easily scaled down for high-throughput screening purposes in multidishes.

Another important aspect for the success of this protocol is that the producer cell line used grows in serum-free medium. Serum is the main source of contaminants in harvested supernatants and is the predominant cost factor for large-scale vector production. Serum supplementation increases the complexity, duration, and cost of downstream processing operations and also raises regulatory concerns due to the risk of introducing adventitious agents. In this chapter, we describe a heparin affinity chromatography procedure that allows for the concentration and purification of lentiviral supernatants produced under serum-free conditions in a single step (16). Unlike traditional ultracentrifugation techniques, chromatography enables fast, efficient, and reproducible separation of viral particles and this purification strategy is scalable (19). In addition, heparin affinity chromatography purification typically results in high recoveries of infective particles because of the mild conditions required for viral vector binding and elution and should be useful for the purification of different vector pseudotypes (20–22).

2. Materials

2.1. Suspension Cell Culture

1. Cell culture capabilities and Biosafety Level II containment (see Note 1).

2. Human embryonic kidney 293 cell line (HEK 293) adapted to grow in suspension culture and serum-free medium (Biotechnology Research Institute, National Research Council of Canada) (see Note 2).

3. Hyclone SFM4 Transfx293™ medium (HyClone Cat. No. SH30860) supplemented with 0.1% Pluronic® F-68 (Invitrogen Cat. No. 24040-032) or any other serum-free medium that supports HEK 293 suspension cell growth and allows for PEI-based transfection (see Note 3).

4. Sterile Dimethyl Sulfoxide (DMSO), cell culture grade.

5. Neubauer hemacytometer and exclusion dye (Trypan Blue 0.1% in PBS or Erythrosin B 25 mg/mL in PBS).

6. Disposable 125-mL polycarbonate Erlenmeyer flasks (Corning Life Science, Cat. No. CLS431143) (see Note 4).

7. Humidified cell culture incubator at 37°C containing 5% CO_2 in air.

8. CO_2 and humidity resistant orbital shaker (see Note 5).

2.2. Plasmid Amplification and Purification

1. *Escherichia coli* DH5α strain (Invitrogen Cat. No. 18265-017) and SURE® strain (Stratagene Cat. No. 200238).

2. CircleGrow agar plates and medium (Qbiogene Cat. No. 3000-122) containing appropriate selective antibiotic (e.g., 100 μg/mL of ampicillin) (Sigma Cat. No. A9518).

3. Maxi/Giga prep plasmid purification kit (Qiagen Cat. No. 12163/12191).

4. Buffer: 10 mM Tris–HCl, 1 mM EDTA (TE buffer), pH 8.

2.3. PEI-Mediated Transient Transfection of Suspension Cells

1. Linear PEI, MW 25,000 (Polysciences Cat. No. 23966). Sterile stock solution: 1 mg/mL of PEI in Milli-Q H_2O, pH 7 (see Note 6).

2. Purified plasmids: Self-inactivating lentiviral transfer vector (pCSII-CMV5-GFPq) (9), Env-protein plasmid (pSVCMVin) coding for the VSV-G protein (23) and third generation packaging plasmids (pMDLg/pRRE#54 and pRSV-Rev) coding for Gag-Pol and Rev sequences, respectively (18).

2.4. Concentration and Purification by Heparin Affinity Chromatography

1. Low-pressure liquid chromatography system (FPLC).

2. Fractogel® EMD heparin (S) gel (Merck) packed into an HR 5/5 column (GE Healthcare) to a final volume of 1 mL (see Note 7).

3. 0.45 μm pore size Acrodisc syringe-mounted filters with HT Tuffryn® polysulfone membrane (Pall Life Sciences).

4. Buffer A: 20 mM Tris–HCl, pH 7.5 filtered and degassed (see Note 8).

5. Buffer B: 20 mM Tris–HCl, 2 M NaCl, pH 7.5 filtered and degassed.

6. Storage buffer: 150 mM NaCl in Milli-Q H_2O, 20% Ethanol filtered and degassed.

7. Regeneration/sanitization buffer: 2 M NaCl in Milli-Q H_2O, 0.1 M NaOH filtered and degassed (see Note 9).

2.5. Titration of Lentiviral Vectors with Suspension Cells

1. Flow cytometry capabilities.

2. HEK 293 cells.

3. Multidish 12 well, Nunclon™ (Nunc Cat. No. 150628).

4. Polybrene stock 8 mg/mL, filter sterilized.

5. Phosphate-buffered saline (PBS), pH 7.4.

6. Formaldehide stock 16% in PBS. Keep at 4°C.

7. FACS tubes.

3. Methods

3.1. Suspension Cell Culture Procedures

1. The cell line used for lentivirus vector production and titration is the HEK 293 suspension cell line. This cell line grows in serum-free HyClone SFM4 Transfx293™ medium supplemented with 0.1% Pluronic® F-68.

2. HEK 293 cells are preserved in liquid nitrogen in cryovials. Cells must be frozen while in exponential cell growth. Each vial should contain 1 mL of cell suspension with 5×10^6 and 5×10^7 viable cells/mL in freezing medium (90% serum-free medium and 10% DMSO). Cryopreserved cells are recovered by rapidly thawing the vial in a 37°C water bath. The entire content of the vial is transferred into a 15 mL tube containing 9 mL of prewarmed serum-free medium and centrifuge at $350 \times g$ for 5–10 min. Resuspend cell pellet in 10 mL of medium, transfer cell suspension to a 10 cm dish and incubated 24 h in static conditions before transferring into shake flasks. A new vial is thawed every ~3 months.

3. Maintain suspension cell cultures in 125 mL shake flasks containing 20 mL of cell suspension. Flasks are placed on an orbital shaker platform rotating at 110–120 rpm incubated at 37°C in a humidified atmosphere containing 5% CO_2 in air (Fig. 1). Cell count and viability are measured daily using a Neubauer hemocytometer and an exclusion dye.

4. Subculture cells while in exponential growth phase, i.e., when the cell density reaches $1–2 \times 10^6$ cells/mL, dilute the suspension cell culture in fresh medium down to 3×10^5 cells/mL (typically 2 or 3 times a week). HEK 293 cell cultures should grow as single-cell suspensions (see Note 10). Healthy HEK 293 cells show viability over 90% at all times, a doubling time of 24 h and can achieve high cell densities ($4–5 \times 10^6$ viable cells/mL).

3.2. Plasmid Amplification and Purification

1. Transform *E. coli* competent cells with the different plasmids required for lentiviral vector production by standard methods (see Note 11). Plate on agar medium with antibiotic and incubate overnight at 37°C.

2. Use freshly transformed cell colonies for plasmid amplification. Grow bacterial cell cultures in medium containing antibiotic during 24 h (see Note 12).

3. Purify plasmids according to the manufacturer's instructions. Check the plasmid purity and concentration by UV absorbance at 260 and 280 nm and DNA integrity by agarose gel electrophoresis (see Note 13). Adjust plasmid concentration to ~1 mg/mL of DNA in TE buffer, aliquot and freeze at –20°C.

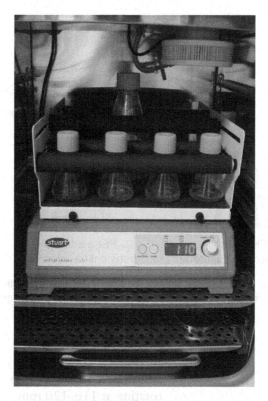

Fig. 1. Suspension cell cultures. Suspension cell cultures grown in shake flasks and multidishes agitated on an orbital shaker platform placed in the cell culture incubator

3.3. PEI-Mediated Transient Transfection of Suspension Cells (Fig. 2)

1. The day before the experiment, split HEK 293 cells to a density of 5×10^5 cells/mL (see Note 14). Grow cultures overnight and determine cell density and viability. Cell density at transfection should be $\sim 1 \times 10^6$ cells/mL and cell culture viability over 95%.

2. Transfer 18 mL of cell suspension into a new disposable 125 mL flask and return the cell culture to the incubator until transfection (see Note 15). Warm-up the culture medium to 25–37°C. Thaw the DNA and the PEI reagent at room temperature.

3. Transfer 2 mL of transfection medium into a 15 mL sterile tube and add a total of 20 μg of DNA with a VSV-G: Gag-Pol: Rev: transfer vector plasmid mass ratio of 1:1:1:2. This is equivalent to 4 μg of each plasmid, except for the transfer vector, which is added in excess (8 μg). Vortex gently (see Note 16).

4. Add 60 μL of stock PEI solution (60 μg). Vortex gently.

5. Incubate the transfection mixture for 15 min at room temperature to allow PEI–DNA complex formation.

6. Transfect cells by adding the whole PEI–DNA mixture to the culture and swirl the flask (see Notes 17 and 18).

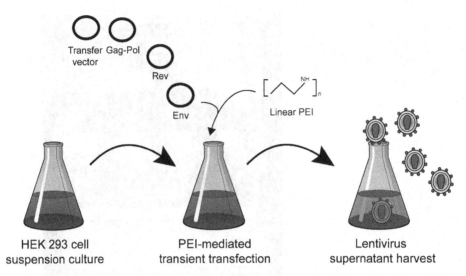

Fig. 2. Lentiviral vector production strategy. The scheme depicts the production of lentiviral vector particles by PEI-mediated transient transfection of HEK 293 suspension growing cells using a third generation lentiviral plasmid system

7. Incubate flasks in a 37°C incubator containing a humidified atmosphere of 5% CO_2 in air on an orbital shaker platform rotating at 110–120 rpm.

8. Harvest supernatants containing infective lentiviral particles daily by centrifugation of the cell suspension in a 50 mL sterile tube at $350 \times g$ during 5-10 min. Store the supernatant at 4°C and resuspend the cell pellet in 20 mL of fresh medium. Return the cell suspension to the incubator in the same shake flask and incubate for another 24 h.

9. Pool supernatants containing the highest lentiviral vector titers. Under the conditions described herein, we obtain sustained high titers from day 3 to 5 posttransfection. Filter the 60 mL of pooled supernatant through 0.45 µm membranes using an Acrodisc syringe-mounted filter. Store vector stocks at –80°C or keep at 4°C overnight for subsequent chromatography purification.

3.4. Concentration and Purification by Heparin Affinity Chromatography

1. Start FPLC system and rinse with Milli-Q H_2O.

2. Install the 1-mL Fractogel® heparin column and remove the storage buffer with 10 column volumes (CV) of Milli-Q H_2O at 0.3 mL/min.

3. Rinse lines A and B with the corresponding buffers and equilibrate the column with 10 CV of binding/wash buffer containing 150 mM NaCl (7.5% buffer B) at 0.5 mL/min (153 cm/h linear flow rate) (see Note 19).

4. Thaw lentivirus supernatants using water bath at 37°C (if required).

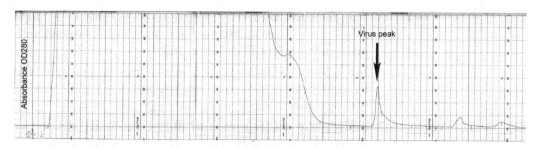

Fig. 3. Typical heparin affinity chromatography elution profile. Clarified lentiviral vector supernatant (35 mL) was loaded onto a 1 mL Fractogel ® EMD Heparin (S) column. The virions are eluted by addition of 350 mM NaCl into the mobile phase using a previously designed elution strategy for the purification of gamma-retroviral particles (20–22)

5. Refilter the sample using an Acrodisc syringe-mounted filter with a pore size of 0.45 μm if the supernatants were frozen. Aliquot starting material samples for analyses.

6. Monitor the UV absorbance at 280 nm. When a stable baseline is achieved, load lentivirus supernatant.

7. Apply a step-wise gradient elution strategy that includes a wash step at 150 mM NaCl (7.5% buffer B, 15 CV) to remove the bulk of contaminating proteins, followed by virus elution at 350 mM NaCl (17.5% buffer B, 13 CV) and a final high-stringency wash step at 1.2 M NaCl (60% buffer B, 6.5 CV) to remove tightly bound contaminants. The process is carried out at room temperature and 0.5 mL/min.

8. The virus particles elute in a defined peak at 350 mM NaCl (Fig. 3). Pool virus-containing fractions and aliquot for analyses.

9. After each run, re-equilibrate the column with binding buffer (10 CV) at 0.5 mL/min or store the column in storage buffer (10 CV) at 0.3 mL/min.

10. Store purified samples at –80°C along with the previously aliquoted samples.

3.5. Titration of Lentiviral Vectors by Flow Cytometry Using Suspension Growing Target Cells (see Note 20)

1. Count exponentially growing HEK 293 cells using a Neubauer hemacytometer and an exclusion dye. Warm-up the culture medium to 25–37°C.

2. Dilute cells in sterile 50 mL tubes to a density of 5.5×10^5 cells/mL with fresh culture medium. Add polybrene directly to this cell suspension to achieve a final concentration of 8 μg/mL.

3. Plate cells in 12-well plates (0.9 mL/well) and transduce with 0.1 mL of virus sample (nondiluted or diluted in fresh culture medium).

4. Incubate plates at 37°C on the orbital shaker (110–120 rpm) for 48 h at 37°C in a humidified atmosphere containing 5% CO_2 in air.

5. Transfer transduced cell suspensions to 1.5 mL Eppendorf tubes, centrifuge at $350 \times g$ for 5 min and resuspend cell pellet in 0.5 mL of PBS. Vortex mildly.

6. Fix cells by addition of 0.5 mL of freshly prepared 4% formaldehyde in PBS and incubate samples for 30 min at 4°C (see Note 21).

7. Transfer the cell suspension to an FACS tube and determine the percentage of transduced GFP+ cells by flow cytometry analysis (see Note 22).

8. Viral titers are calculated with the following formula:
   Titer $(IVP/mL) = (\%GFP+ \text{ cells}/100) \times (\text{Number of cells at time of exposure}) \times (\text{virus dilution factor})/(\text{sample volume})$; being the number of cells at time of exposure $\sim 5 \times 10^5$ cells and the sample volume 1 mL.

4. Notes

1. Lentiviral vectors must be handled under appropriate biosafety containment by trained personnel using biological safety cabinets and following the guidelines specified by the institution, where the experiments are conducted. The nature of the transgene must be taken into account when establishing biosafety requirements. The work described herein has been performed in biosafety level-2 laboratories.

2. The cell line used for lentivirus vector production and titration is the HEK 293. The cell line has been developed by NRC for serum-free suspension cell culture (US Patent # 6,210,922) and deposited at ATCC (CRL-12585). A cGMP master cell bank has been manufactured for clinical applications. In the previous work, the HEK 293 EBNA-1 cell line (clone 6E) stably expressing a truncated functional form of the EBV nuclear antigen-1 was used giving similar results. These cell lines have been selected due to their ability to grow in suspension under serum-free conditions and produce high amounts of recombinant proteins upon transient transfection. Other highly transfectable suspension and serum-free adapted cell lines could be employed for vector production, although vector yields may be lower.

3. Suspension culture medium is typically supplemented with 0.1% Pluronic® F-68 to avoid shear stress in culture. Note that some commercially available suspension culture media contain anticlumping agents (e.g., dextran sulfate, heparin) than can inhibit PEI-mediated transfection.

4. This protocol describes lentiviral vector production at small scale using 125 mL shake flasks containing 20 mL of cell culture.

This procedure can be easily scaled up to 650 mL cell culture in a 2-L shake flask. The cell culture working volume should range between 16 and 33% of the total shake flask volume to allow adequate culture oxygenation. Larger volumes of vector supernatant can be produced in bioreactors. The process described here has been successfully scaled up from shake flasks to a 3-L bioreactor (16). The method can also be scaled down for high-throughput screening purposes. In this case, 6-well plates with a final cell suspension culture volume of 2 mL are recommended.

5. Ensure that the orbital shaker can function under humid conditions. Do not stop the orbital shaker while it is inside the humidified incubator for extended period of time as this may cause the shaker to seize.

6. PEI can be dissolved by lowering the pH of the solution below 2 by drop wise addition of concentrated HCl. Stir until PEI is completely dissolved and then neutralize with 10 M NaOH. Adjust volume to obtain a final concentration of 1 mg/mL of PEI. Filter-sterilize the solution through a 0.22 μm membrane. Aliquot and keep stock solutions at –20°C.

7. It should be noted that some chromatographic columns, such as the HR5/5, are provided with top and bottom filters designed for protein purification purposes. To avoid losses of viral particles trapped on the filter, replace the original filters with a suitable filter (i.e., 10 μm mesh).

8. It is essential that all buffers used for chromatography are freshly filtered (0.45 μm) and degassed to protect the column. Samples should also be filtered through 0.45 μm membranes before being loaded into the column.

9. It is recommended that after 3–4 purification runs the columns are regenerated/ sanitized. This is normally performed by washing with 2–3 column volumes of regeneration/sanitization solution followed by an extensive wash with Milli-Q H$_2$O to bring back the column to neutral pH. Monitor the column effluent pH before beginning the next run.

10. Suspension cells sometimes grow as small clusters. Vigorous vortexing may be required at each subculture step for a number of passages until the cultures grow predominantly as single-cell suspensions. Extensive clumping affects cell growth and transfection efficiency.

11. To minimize the frequency of homologous recombination between LTR regions in *E. coli*, it is recommended that lentiviral transfer vector plasmids be propagated using Rec-bacterial strains (e.g., SURE® strain).

12. We routinely obtained approximately 10 g of cell pellet for every 0.5 L of bacterial cell culture and routinely recovered ~10 mg of pure plasmid.

13. For optimal transfection efficiency, high-quality plasmid stocks should be used. An OD 260/280 ratio over 1.8 is preferred. Plasmids should be verified by agarose gel electrophoresis to confirm that the preparation is mostly in supercoiled form and free of RNA contamination. Use a supercoiled ladder to verify plasmid sizes.

14. Cell culture should be diluted in fresh medium a day or 2 before transfection and allowed to grow to the desired density for optimal transfection. Avoid centrifuging cells the day of transfection as this can have a negative effect on transfection efficiency.

15. The protocol describes production of lentiviral particles by transient transfection of 20 mL suspension cultures. The production scale can be adjusted to fit other experimental requirements as outlined in Note 4.

16. The method has been successfully tested with different combination and number of plasmid constructs coding for lentivirus vector. Therefore, the method can be easily adapted for lentivirus production with one preferred combination of plasmids after determination of optimal plasmid ratios.

17. The final DNA concentration in culture medium is 1 μg/mL and the DNA to PEI mass ratio should be in the range of 1:3. These parameters should be kept constant for lentivirus vector production regardless of the production scale.

18. The lentiviral transfer vector used in our studies carries a fluorescent marker (GFP) that allows for assessment of the transfection efficiency using fluorescence microscopy or flow cytometry.

19. It is recommended to perform an acetone test to verify the packing efficiency and column integrity and a blank test prior to the run with running buffer.

20. This method can be used to titer lentiviral vectors coding for GFP. It should be noted that the titration assay is basically the same as the traditional reporter gene expression assay with the exception that instead of using adherent target cells we have successfully employed suspension-growing target cells. No overnight cell attachment and trypsinization steps are required, thus, reducing the time and greatly simplifying the titration assay.

21. Fixed samples can be stored in the dark at 4°C overnight. Formaldehyde has the double function of fixing the cells and inactivating the viral particles. Samples can be taken outside the contained laboratory for flow cytometry analysis after this step.

22. Samples from a given experiment should be analyzed in a single titration assay to avoid inter-assay variability and all samples should be processed in duplicate to assess intra-assay variability. Only virus dilutions that result in %GFP+ cell values ranging from 3 to 25% should be considered for analysis.

References

1. Naldini, L., Blomer, U., Gallay, P., Ory, P., Mulligan, R., Gage, F.H., et al. (1996). In vivo gene delivery and stable transduction of nondividing cells by a lentiviral vector. *Science* 272, 263-7.

2. Kafri, T., van Praag, H., Ouyang, L., Gage, F.H., and Verma, I.M. (1999). A packaging cell line for lentivirus vectors. *J Virol* 73, 576-84.

3. Lesch, H.P., Turpeinen, S., Niskanen, E.A., Mähönen, A.J., Airenne, K.J., and Ylä-Herttuala, S. (2008). Generation of lentivirus vectors using recombinant baculoviruses. *Gene Ther* 15, 1280-6.

4. Bartz, S.R., Rogel, M.E., and Emerman, M. (1996). Human immunodeficiency virus type 1 cell cycle control: Vpr is cytostatic and mediates G2 accumulation by a mechanism which differs from DNA damage checkpoint control. *J Virol* 70, 2324-31.

5. Konvalinka, J., Litterst, M.A., Welker, R., Kottler, H., Rippmann, F., Heuser, A.M., et al. (1995). An active-site mutation in the human immunodeficiency virus type 1 proteinase (PR) causes reduced PR activity and loss of PR-mediated cytotoxicity without apparent effect on virus maturation and infectivity. *J Virol* 69, 7180-6.

6. Li, C.J., Friedman, D.J., Wang, C., Metelev, V., and Pardee, A.B. (1995). Induction of apoptosis in uninfected lymphocytes by HIV-1 Tat protein. *Science* 268, 429-31.

7. Miyazaki, Y., Takamatsu, T., Nosaka, T., Fujita, S., Martin, T.E., and Hatanaka, M. (1995). The cytotoxicity of human immunodeficiency virus type 1 Rev: implications for its interaction with the nucleolar protein B23. *Exp Cell Res* 219, 93-101.

8. Burns, J.C., Friedmann, T., Driever, W., Burrascano, M., and Yee, J.K. (1993). Vesicular stomatitis virus G glycoprotein pseudotyped retroviral vectors: concentration to very high titer and efficient gene transfer into mammalian and nonmammalian cells. *Proc Natl Acad Sci U S A* 90, 8033-7.

9. Broussau, S., Jabbour, N., Lachapelle, G., Durocher, Y., Tom, R., Transfiguracion, J., et al. (2008). Inducible packaging cells for large-scale production of lentiviral vectors in serum-free suspension culture. *Mol Ther* 16, 500-7.

10. Sena-Esteves, M., Tebbets, J.C., Steffens, S., Crombleholme, T., and Flake, A.W. (2004). Optimized large-scale production of high titer lentivirus vector pseudotypes. *J Virol Methods* 122, 131-9.

11. Follenzi, A. and Naldini, L. (2002). HIV-based vectors. Preparation and use. *Methods Mol Med* 69, 259-74.

12. Tiscornia, G., Singer, O., and Verma, I.M. (2006). Production and purification of lentiviral vectors. *Nat Protoc* 1, 241-5.

13. Salmon, P. and Trono, D. (2007). Production and titration of lentiviral vectors. *Curr Protoc Hum Genet* Chapter 12, Unit 12 10.

14. Tonini, T., Claudio, P.P., Giordano, A., and Romano, G. (2004). Transient production of retroviral- and lentiviral-based vectors for the transduction of Mammalian cells. *Methods Mol Biol* 285, 141-8.

15. Chang, L.J. and Zaiss, A.K. (2002). Lentiviral vectors. Preparation and use. *Methods Mol Med* 69, 303-18.

16. Segura, M.M., Garnier, A., Durocher, Y., Coelho, H., and Kamen, A. (2007). Production of lentiviral vectors by large-scale transient transfection of suspension cultures and affinity chromatography purification. *Biotechnol Bioeng* 98, 789-99.

17. Durocher, Y., Perret, S., and Kamen, A. (2002). High-level and high-throughput recombinant protein production by transient transfection of suspension-growing human 293-EBNA1 cells. *Nucleic Acids Res* 30, E9.

18. Dull, T., Zufferey, R., Kelly, M., Mandel, R.J., Nguyen, M., Trono, D., et al. (1998). A third-generation lentivirus vector with a conditional packaging system. *J Virol* 72, 8463-71.

19. Segura, M.M., Kamen, A., and Garnier, A. (2006). Downstream processing of oncoretroviral and lentiviral gene therapy vectors. *Biotechnol Adv* 24, 321-37.

20. Segura, M.M., Kamen, A., and Garnier, A. (2008). Purification of retrovirus particles using heparin affinity chromatography. *Methods Mol Biol* 434, 1–11.

21. Segura, M.M., Kamen, A., Lavoie, M.C., and Garnier, A. (2007). Exploiting heparin-binding properties of MoMLV-based retroviral vectors for affinity chromatography. *J Chromatogr B Analyt Technol Biomed Life Sci* 846, 124–31.

22. Segura, M.M., Kamen, A., Trudel, P., and Garnier, A. (2005). A novel purification strategy for retrovirus gene therapy vectors using heparin affinity chromatography. *Biotechnol Bioeng* 90, 391–404.

23. Kobinger, G.P., Weiner, D.J., Yu, Q.C., and Wilson, J.M. (2001). Filovirus-pseudotyped lentiviral vector can efficiently and stably transduce airway epithelia in vivo. *Nat Biotechnol* 19, 225–30.

Chapter 3

Flexibility in Cell Targeting by Pseudotyping Lentiviral Vectors

Daniela Bischof and Kenneth Cornetta

Abstract

Lentiviral vectors have become an important research tool and have just entered into clinical trials. As wild-type lentiviruses engage specific receptors that have limited tropism, most investigators have replaced the endogenous envelope glycoprotein with an alternative envelope. Such pseudotyped vectors have the potential to infect a wide variety of cell types and species. Alternatively, selection of certain viral envelope glycoproteins may also facilitate cell targeting to enhance directed gene transfer. We describe the method for generating pseudotyped vector and provide information regarding available pseudotypes and their respective target tissues.

Key words: Lentiviral vectors, Pseudotyping, Vesicular stomatitis Virus G Protein, Glycoprotein, Tropism

1. Introduction

Many investigators have utilized the HIV-1 virus as the basis for a lentiviral vector construct. The transgene vector is created by utilizing the viral genomic regions required for integration and packaging (Long Terminal Repeats or LTRs and packaging sequence), while deleting the viral protein coding sequences (*gag*, *pol*, and *env*). Vectors generally delete HIV-1 accessory genes to improve the safety profile, a modification that also maintains the ability of these vectors to transduce nondividing cells (1–4). Many investigators have also utilized self-inactivating vectors that remove the enhancer and promoter functions from the viral LTRs as another means of increasing safety (5, 6). Inserted into this vector "backbone" are the desired transgene(s) of interest along with the regulatory sequences that drive transgene expression.

Maurizio Federico (ed.), *Lentivirus Gene Engineering Protocols,* Methods in Molecular Biology, vol. 614, DOI 10.1007/978-1-60761-533-0_3, © Humana Press, a part of Springer Science+Business Media, LLC 2010

In addition to HIV-1, lentivirus vectors have been engineered with HIV-2 (7, 8), Simian Immunodeficiency Virus (9, 10), Feline Immunodeficiency Virus (11, 12), the Equine Immunodeficiency Virus (13), Caprine Arthritis-Encephalitis Virus (14), Bovine Immunodeficiency Virus (15), Jembrana Disease Virus (16), and Visna Virus (17).

An important modification that must be made when generating lentiviral vectors is the replacement of the native HIV-1 envelope glycoprotein with an alternative envelope. This is required because the wild-type HIV envelope glycoprotein utilizes CD4 as its major receptor, a cell surface molecule found predominately on T cells and macrophages. Therefore, to allow gene transfer into cells that do not express CD4, the vector must be pseudotyped with an alternative envelope.

Pseudotyping was first described when investigators noted that host cells infected with two enveloped viruses generated progeny, with both the native envelope and the coinfected viral envelope. For example, cells infected with the Vesicular Stomatitis Virus (VSV) and a retrovirus can produce VSV particles expressing the VSV envelope and the Moloney Leukemia virus (MLV) and Avian Myeloblastosis Virus (AMV) envelopes (18, 19). Similarly, pseudotyping has been shown for cells infected with parainfluenza and VSV (20). Pseudotyping has long been used in the context of retroviral gene transfer. Initial constructs were based on the MLV, a gamma retrovirus expressing an envelope glycoprotein that facilitates infection of rodent cells, but not primate cells. Pseudotyping with the murine 4070A and other viral envelopes have allowed these vectors to infect the majority of mammalian cells (see for a review (21)).

Lentiviruses are membrane-bound virions that acquire their membrane immediately prior to being released from the host cell. The envelope of HIV-1 is processed prior to incorporation into virions; the envelope is synthesized as a precursor (gp160), which is cleaved by a host protease resulting in the surface (gp120) and transmembrane (gp41) subunits (22). The envelope is glycosylated and accumulates in lipid rafts located within the plasma and intracellular membranes. While the envelope glycoprotein is important for tropism, it is not required for virion budding (23), an important property of the lentivirus that facilitates pseudotyping. Furthermore, envelopes from viruses, such as VSV and influenza virus, are also known to accumulate in lipid rafts and bud from the plasma membrane, further facilitating pseudotyping (24).

The primary function of envelope glycoprotein is to fuse the viral envelope and the host plasma membrane. In the case of HIV-1, infection of target cells involves interaction of the major receptor, CD4, and one of two co-receptors either CXCR4 or CCR5. The virus–plasma membrane fusion that follows is mediated by gp41 (25, 26).

The observation that the VSV glycoprotein (VSV-G) can efficiently pseudotype HIV-1 virions was a major advance in the development of lentiviral gene transfer (27–29). Viral entry of RNA viruses via envelope glycoproteins can occur via a pH-dependent (receptor mediated endocytosis) or a pH-independent mechanism (direct entry). In pH-dependent host cell entry, the virus is endocytosed, and the internalized vesicle is acidified in the cytoplasm after being pinched off from the cell membrane. This drop in pH facilitates fusion of the viral envelope and host endosome, thus releasing the viral genome into the cytoplasm (30). The envelope glycoproteins of VSV, Ross River virus (RRV), human foamy virus, and Jaagsiekte Sheep retrovirus (JSR) are involved in pH-dependent cell entry (31–33). The pH-independent entry mechanism involves the direct fusion of the viral envelope with the cell membrane, inducing the release of viral ribonucleoprotein complexes into the cytoplasm (30). HIV-1, RD114, and measles virus exhibit such direct viral entry (34–36). VSV-G pseudotyping imparts virion stability, thus allowing concentration by ultracentrifugation (37). While the receptor for VSV-G envelope has not been identified, it allows the delivery of lentiviral particles in the cytoplasm of a wide range of species and cell types. The majority of investigators, conducting lentiviral gene transfer, utilize this envelope for pseudotyping.

While the broad tropism of VSV-G is desirable for many applications, it can be problematic in some instances. The fusogenic nature of the VSV-G glycoprotein has proven problematic when generating packaging cell lines. The ability to infect almost any cell is not an advantage when tissue specificity or cell targeting is required. For these reasons, investigators are exploring alternative envelopes.

Envelope glycoproteins from other retroviruses have been used to pseudotype lentiviral vectors. The native envelopes from human foamy virus, Gibbon Ape Leukemia Virus, and the feline RD114 virus can all pseudotype HIV-1, but the infectious titers of the resultant material is low (38–42). Di Nunzio et al. have shown that improved yield of infectious particles can be obtained by truncating and fusing the extracellular and TM domains of RD114 to the cytoplasmic tail of MLV envelope (43). This modification has led to gene transfer that has been reported by some investigators to be approaching that of VSV-G, although there are conflicting data as to the optimal retroviral envelope for pseudotyping lentiviral vectors (42, 44, 45). This may relate to the varied transduction protocols and target cells used by different investigators.

Investigators have utilized lentiviral vectors pseudotyped with alpha-viral envelopes to provide efficient gene transfer (46–48). Kang et al. compared Ross River virus (RRV) Envelope and VSV-G pseudotyped FIV and noted increased transduction of murine hepatocytes in vivo, and also reported efficient transduction of neuroglial cells when injected into the brain (49). Kahl et al. noted preferentially low transduction of RRV pseudotyped HIV-1-based lentiviral vectors into cells of hematopoietic origin. This finding could prove advantageous when trying to obtain organ-specific gene transfer while avoiding transfer into blood cell precursors and their progeny (50). Cell targeting has long been a goal of viral gene therapy, although prior modifications that have increased specificity have often resulted in decreased transduction efficiencies. Recently, Funke et al. have obtained specificity and high transduction by altering the measles virus glycoprotein envelope to preferentially infect CD20+ cells as a means of targeting lymphoid tissue (51).

Lentiviral vectors targeting the lung have shown interesting specificity when vectors are pseudotyped with the Jaagsiekte Sheep retrovirus (JSR) envelope. An FIV vector pseudotyped with this envelope provided efficient transduction of type II cells, but little gene transfer was seen in airway epithelia, the normal target for the JSR virus (52). Overexpression of the JSR receptor in airway epithelia did not improve gene transfer, suggesting restriction was due to mechanisms unrelated to the envelope. This suggests that the wild-type tropism for an envelope may not be directly transferable to lentiviral particles. In general, it will be important to test pseudotyped lentiviral vectors in the specific cell or tissue that is the intended target and explore other envelopes if the desired level of gene transfer is not obtained.

In this chapter, we will present a method for generating pseudotyped lentiviral vector. We will also discuss points to consider when selecting an envelope for pseudotyping lentiviral vectors. While a complete discussion of all potential pseudotypes is beyond the scope of this paper, we provide a list of viruses that have successfully pseudotyped lentiviral vectors (Table 1). We also provide a list of lentiviral pseudotypes and their application in various organ and tissue targets (Table 2).

2. Materials

1. Biologic safety cabinet (BSC), tissue culture incubator, centrifuge, and <-70°C freezer.
2. D10 medium for cell expansion: Dulbecco's Modified Eagle Medium (D-MEM) supplemented with 2 mM L-glutamine and 1 mM sodium pyruvate (GIBCO/Invitrogen) and containing 10% fetal bovine serum (HyClone) (see Note 1).
3. For cell expansion and transfection: 75 cm² tissue culture flasks, calibrated disposable pipettes (various sizes), centrifuge tubes (15 and 50 mL, BD Falcon™ or Corning Inc.).

Table 1
Viral envelopes used to pseudotype lentiviral particles

Virus family	Pseudotyping glycoprotein	Lentivirus pseudotyped	References
Rhabdoviridae	VSVG	HIV-1	(28, 29, 55–58)
		HIV-2	(59)
		SIV	(60)
		EIAV	(61–63)
		FIV	(11)
		CAEV	(14)
		BIV	(15)
		JDV	(16)
		VV	(17)
	Rabies	HIV-1	(55)
		EIAV	(61, 62)
	Mokola	HIV-1	(55, 56, 58)
		EIAV	(62)
		SIV	(60)
Orthomyxoviridae	Influenza	HIV-1	(64)
		EIAV	(64)
		SIV	(60, 65)
Togaviridae	RRV	HIV-1	(54, 66)
		FIV	(67)
	SFV	HIV-1	(54)
	EEV	HIV-1	(68, 69)
	Sindbis virus	HIV-1	(70)
Filoviridae	Ebola	HIV-1	(58)
		FIV	(71)
	Marburg	FIV	(71)
Retroviridae	GALV and /TR	HIV-1	(72, 73)
	Ampho	HIV-1	(57, 72)
		SIV	(60)
	MuLV	HIV-1	(58)

(continued)

**Table 1
(continued)**

Virus family	Pseudotyping glycoprotein	Lentivirus pseudotyped	References
		EIAV	(61)
	RD114 and /TR	HIV-1	(57, 72, 74, 75)
		SIV	(60)
	Eco	HIV-1	(57, 76)
	HTLV-1	HIV-1	(77)
	JSRV	HIV-1	(78)
	ALSV	HIV-1	(79)
Paramyxioviridae	MV	HIV-1	(80)
	HPIV-3	HIV-1	(81)
	Sendai virus	HIV-1	(82)
		SIV	(83)
Arenaviridae	LCMV	HIV-1	(58, 84)
		EIAV	(62)
		SIV	(60)
Flaviviridae	HCV	HIV-1	(85)
Baculoviridae	GP64	HIV-1	(86)

4. For cell passage: Dulbecco's phosphate buffered saline without calcium or magnesium (PBS) and Trypsin–EDTA.

5. Cell Line: HEK293T cells (available through ATCC, www.atcc.org).

6. Plasmids: lentiviral vector plasmid containing transgene of interest (investigators choice); packaging plasmid (containing *gag* and *pol* genes); Rev plasmid if using third-generation lentiviral vector system; envelope plasmid (containing the desired envelope pseudotype). For this discussion, we will utilize the third-generation HIV-1 based system as an example. The plasmid amounts are given per 75 cm^2 flask. Concentrations for the pMDL packaging plasmid containing HIV-1 *gag/pol* and the pRSV/Rev plasmid, described in Dull et al. (53), are 6.6 µg and 3.3 µg, respectively. For envelope plasmids, we utilized 5.1 µg for VSV-G pseudotyping and 6.3 µg for RRV envelope when using plasmids previously described in Kahl et al. (pCI-MD.G or pCI-RRV) (54). For a GFP transgene

Table 2
Tissue and organ targets of pseudotyped lentiviral vectors

Target tissue/organ (various species)	Pseudotyping glycoprotein/ lentivirus	References
Lung	VEEV/HIV-1	(68)
	WEEV/HIV-1	(69)
	JSRV/HIV-1	(78)
	RRV/HIV-1	(54)
	SFV/HIV-1	(54)
	Influenza/HIV-1	(64, 87)
	Influenza/EIAV	(64)
	Marburg/FIV	(71)
	Ebola/FIV	(71)
	Ebola/HIV-1	(87)
	Sendai/SIV	(83)
	VSVG/HIV-1	(87)
	VSVG/EIAV	(63)
CNS	VEEV/HIV-1	(68)
	WEEV/HIV-1	(69)
	VSVG/EIAV	(62)
	Mokola/EIAV	(62)
	Rabies/EIAV	(62, 88, 89)
	Mokola/HIV-1	(56, 58)
	VSVG/HIV-1	(56, 58, 90)
	VSVG/HIV-2	(59)
	VSVG/BIV	(15)
	LCMV/FIV	(91)
	LCMV/HIV-1	(92)
Lymphocytes hematopoietic cells	VEEV/HIV-1	(68)
	MV/HIV-1	(80)
	RD114/SIV	(65)
	RD114/HIV-1	(57, 75, 93)

(continued)

Table 2
(continued)

Target tissue/organ (various species)	Pseudotyping glycoprotein/ lentivirus	References
	Eco/HIV-1	(57, 76)
	Ampho/HIV-1	(57, 93)
	VSVG/HIV-1	(57, 74)
	VSVG/HIV-2	(59)
	VSVG/BIV	(15)
	HTLV-1/HIV-1	(77)
	GALV/HIV-1	(93)
Fibroblasts	VEEV/HIV-1	(68)
	Ampho/EIAV	(61)
	Eco/EIAV	(61)
	VSVG/EIAV	(61)
	Rabies/HIV-1	(55)
	Mokola/HIV-1	(55)
	ALSV/HIV-1	(79)
	JSRV/HIV-1	(78)
	Ampho/HIV-1	(72)
	VSVG/HIV-1	(72)
	VSVG/FIV	(11)
Kidney	WEEV/HIV-1	(69)
	Ampho/EIAV	(61)
	VSVG/EIAV	(61)
	HPIV-3/HIV-1	(81)
	GP64/HIV-1	(86)
	RRV/HIV-1	(54)
	SFV/HIV-1	(54)
	Sendai/SIV	(83)
	VSVG/HIV-1	(72)
	VSVG/FIV	(11)
	GALV/HIV-1	(72)

(continued)

Table 2
(continued)

Target tissue/organ (various species)	Pseudotyping glycoprotein/ lentivirus	References
	RD114/HIV-1	(72)
Bone (osteosarcoma)	WEEV/HIV-1	(69)
	VSVG/HIV-1	(69)
	Ampho /EIAV	(61)
	VSVG/EIAV	(61)
	GP64/HIV-1	(86)
	RD114/HIV-1	(75)
	VSVG/FIV	(11)
	VSVG/BIV	(15)
Muscle	Rabies/EIAV	(62)
	VSVG/BIV	(15)
	Mokola/HIV-1?	(94)
	Ebola/HIV-1	(94)
Eye	VSVG/SIV	(60)
	VSVG/HIV-1	(95)
	Mokola/SIV	(60)
	Mokola/HIV-1	(95)
	MuLV/SIV	(60)
	Influenza/SIV	(60)
	LCMV/SIV	(60)
Liver	HCV/HIV-1	(85)
	RRV/FIV	(67)
	Sendai/HIV-1	(82)
	Sendai/SIV	(83)
	LCMV/HIV-1	(96)
Dendritic cells	RRV/HIV-1	(66)
Pancreas (Islet cells)	LCMV/HIV-1	(84)

plasmid, we utilized 13.2 μg, but optimization will be required depending on the transgene vector selected (see Note 1).

7. Transfection. Reagents formulated in house or purchased (e.g., Profection Kit, Promega Inc. Madison, WI) and used as per the manufacturer's instructions.

8. If concentration is desired, use ultracentrifuge such as Beckman Optima XL-100 k ultracentrifuge using a 45-Ti fixed-angle rotor, with Beckman Quick-Seal ultracentrifuge tube (Beckman Coulter, Fullerton, CA).

3. Methods

Generation of lentiviral vectors requires cell lines in addition to recombinant plasmids expressing viral genes. Many institutions require review and prior approval of this work. Guidelines vary by country and by institution. At a minimum, work should be performed in Biologic Safety Cabinets with personnel wearing lab coats, eye protection, and gloves.

3.1. Production Using Transient Transfection Method

1. Thaw a vial of HEK293T cells and place in a 75 cm² flask with 12 mL of D10 medium. Remove medium and replace with fresh D10 12–36 h after thaw. Continue to observe the cells daily. When nearing confluence the cells should be split. For the cell passage, the medium should be removed and the monolayer rinsed with PBS, which is then replaced with Trypsin-EDTA (approximately 2 mL per 75 cm²). Cultures should be returned to 37°C for 2–5 min, or until cells detach. Centrifuge the cells, remove the media, and resuspend the cell pellet in fresh D10 medium. Subcultures are generally split at a ratio of 1:6 to 1:8, but this may need to be altered depending on the growth characteristics of the cells. Cells should be re-fed at least every 4 days if they are not ready for passage within that time.

2. Expand cells to obtain the desired number of cells for transduction.

3. Day-1: Prepare a single cell suspension of HEK293T cells and plate 5×10^6 cells in a 75 cm² tissue culture flask. Transfection procedure should begin when cells are 70–80% confluent.

4. Day 2: Remove flask from the incubator and aspirate the medium. Add 12 mL fresh D-10 to each flask and return flasks to the incubator.

5. Day 2: Prepare DNA for transfection by diluting plasmids in sterile water to a total volume of 750 μL. Add 93 μL of 2.0 M CaCl₂. Mix gently. Add DNA mix (750 μL) dropwise to 750 μL

Hepes Buffered Saline (HBS). Remove flasks from incubator and add 1.5 mL DNA/HBS suspension to the flask. Alternative methods for transfection may be used (see Note 2).

6. Day 3: Aspirate and discard medium from flasks. Rinse cells with 5 mL of PBS then aspirate.

7. Day 3: Add 12 mL medium to each flask and return to the incubator. Serum-free media may be used in place of D10 if investigators want to minimize serum in their final product (see Note 3).

8. Day 4: After 20–24 h, remove supernatant from flask. Filter through a 0.45 -μm syringe filter to remove cell debris. Freeze in aliquots based on anticipated needs and store at <–70°C. Additional harvests may be possible (see Notes 4 and 5).

9. If concentration is desired, ultracentrifugation can efficiently concentrate vector. *Prior to freezing*, 94 mL of supernatant can be loaded into the ultracentrifuge tubes, and spun at $50,000 \times g$ for 1 h. Pellets should be resuspended in 1 mL of D-10 or desired medium and stored at <–70°C (see Note 6).

4. Notes

1. The passage number and growth rate of HEK293T cells can affect transfection efficiency. Some optimization of the plasmid concentrations and plating cell density is normally required when seeking to maximize the yield of infectious vector particles. The amounts provided here are starting points for this optimization. Also, antibiotics may be added to the media to decrease the potential for contamination during cell expansion.

2. Vector and packaging constructs must be of high quality to obtain efficient transduction and expression. In generating plasmids, use endotoxin-free plasmid isolation kits (e.g., Qiagen Endotoxin-Free Purification kit). Transfection reagents and buffers used to resuspend plasmids should be sterile-filtered prior to use to prevent contamination of the cell culture during transfection. Once the transfection reagents are mixed, they cannot be sterile-filtered, as the precipitate is the active transfection agent.

3. We utilize calcium phosphate for transient transfection because GMP grade material is readily available. Certain lipofection reagents can provide improved transfection and higher titer. Electroporation techniques are another alternative, but difficult to scale up.

4. If serum-free medium is used during harvests, the titer will fall off more rapidly. For transient transfection, the first harvest provides the highest titer, the second harvest titer is

generally 20–50% of the first harvest, and falls further with subsequent harvests.

5. Certain lentiviral vectors, including VSV-G pseudotyped HIV-1 vectors, are stable overnight at 4°C, so the material can be pooled with the second day harvest prior to processing.

6. If ultracentrifugation is to be performed, the transfection process is generally scaled up to provide more product. Multiple 75 cm^2 flasks or larger flasks can be utilized with the appropriate increase in plasmid based on the ratio of the 75 cm^2 flask to the larger flask.

Acknowledgments

The author is supported in part by the Indiana Genomics Initiative (INGEN). Indiana University is the site of the NHLBI Gene Therapy Resources Program (HHSN26820078204) and the NCRR National Gene Vector Biorepository (P40 RR024928).

References

1. Reiser J, Harmison G, Kluepfel-Stahl S, Brady RO, Karlsson S, Schubert M (1996). Transduction of nondividing cells using pseudotyped defective high-titer HIV type 1 particles. *Proc. Natl. Acad. Sci. U.S.A.* **93**, 15266–71.

2. Kim VN, Mitrophanous K, Kingsman SM, Kingsman AJ (1998). Minimal requirement of a lentivirus vector based on human immunodeficiency virus type 1. *J. Virol.* **72**, 811–6.

3. Sutton RE, Wu HTM, Rigg R, Bohnlein E, Brown PO (1998). Human immunodeficiency virus type 1 vectors efficiently transduce huma hematopoietic stem cells. *J. Virol.* **72**, 5781–8.

4. Naldini L, Blomer U, Gallay P, Ory D, Mulligan R, Gage FH, Verma IM, Trono D (1996). In vivo gene delivery an stable transducion of nondividing cells by a lentiviral vector. *Science* **272**, 263–7.

5. Miyoshi H, Blomer U, Takahashi M, Gage FH, Verma IM (1998). Development of a self-inactivating lentivurs vector. *J. Virol.* **72**, 8150–57.

6. Zufferey R, Dull T, Mandel RJ, Bukovsky A, Quiroz D, Naldini L, Trono D (1998). Self-inactivating lentivirus vector for safe and efficiency in vivo gene delivery. *J. Virol.* **72**, 9873–80.

7. Corbeau P, Kraus G, Wong-Staal F (1998). Transduction of human macrophages using a stable HIV-1/HIV-2-derived gene delivery system. *Gene Ther.* **5**, 99–104.

8. Sadaie MR, Zamani M, Whang S, Sistron N, Arya SK (1998). Towards developing HIV-2 lentivirus-based retroviral vectors for gene therapy, dual gene expression in the context of HIV-2 LTR and Tat. *J. Med. Virol.* **54**, 118–28.

9. Negre D, Duisit G, Mangeot PE, Moullier P, Darlix JL, Cosset FL (2002). Lentiviral vectors derived from simian immunodeficiency virus. *Curr. Top. Microbiol. Immunol.* **261**, 53–74.

10. Gilbert JR, Wong-Staal F (2001). HIV-2 and SIV vector systems. *Somat. Cell. Mol. Genet.* **26**, 83–98

11. Poeschla EM, Wong-Staal F, Looney DJ (1998). Efficient transduction of nondividing human cells by feline immunodeficiency virus lentiviral vectors. *Nat. Med.* **4**, 354–7.

12. Johnston J, Power C (1999). Productive infection of human peripheral blood mononuclear cells by feline immunodeficiency virus, implications for vector development. *J. Virol.* **73**, 2491–8.

13. Olsen JC (2001). EIAV, CAEV and other lentivirus vector systems. *Somat. Cell. Mol. Genet.* **26**, 131–45.

14. Mselli-Lakhal L, Favier C, Leung K, Guiguen F, Grezel D, Miossec P, Mornex JF, Narayan O, Querat G, Chebloune Y (2000). Lack of functional receptors is the only barrier that prevents caprine arthritis-encephalitis virus from infecting human cells. *J. Virol.* **74**, 8343–8.

15. Berkowitz R, Ilves H, Lin WY, Eckert K, Coward A, Tamaki S, Veres G, Plavec I (2001). Construction and molecular analysis of gene transfer systems derived from bovine immuno-deficiency virus. *J. Virol.* **75**, 3371–82.

16. Metharom P, Takyar S, Xia HH, Ellem KA, Macmillan J, Shepherd RW, Wilcox GE, Wei MQ (2000). Novel bovine lentiviral vectors based on Jembrana disease virus. *J. Gene Med.* **2**, 176–85.

17. Berkowitz RD, Ilves H, Plavec I, Veres G (2001). Gene transfer systems derived from Visna virus, analysis of virus production and infectivity. *Virology* **279**, 116–29.

18. Zavada J (1972). Pseudotypes of vesicular stomatitis virus with the coat of murine leu-kaemia and of avian myeloblastosis viruses. *J. Gen. Virol.* **15**, 183–91.

19. Huang AS, Besmer P, Chu L, Baltimore D (1973). Growth of pseudotypes of vesicular stomatitis virus with N-tropic murine leuke-mia virus coats in cells resistant to N-tropic viruses. *J. Virol.* **12**, 659–62.

20. Choppin PW, Compans RW (1970). Phenotypic mixing of envelope proteins of the parainfluenza virus SV5 and vesicular stomati-tis virus. *J. Virol.* **5**, 609–16.

21. Miller AD (1990). Retrovirus packaging cells. *Hum. Gene Ther.* **1**, 5–14.

22. Freed EO, Martin MA (1995). The role of human immunodeficiency virus type 1 enve-lope glycoproteins in virus infection. *J. Biol. Chem.* **270**, 23883–6.

23. Beausejour Y, Tremblay MJ (2004). Envelope glycoproteins are not required for insertion of host ICAM-1 into human immunodeficiency virus type 1 and ICAM-1-bearing viruses are still infectious despite a suboptimal level of tri-meric envelope proteins. *Virology* **324**, 165–72.

24. Pickl WF, Pimentel-Muinos FX, Seed B (2001). Lipid rafts and pseudotyping. *J. Virol.* **75**, 7175–83.

25. Dragic T (2001). An overview of the determi-nants of CCR5 and CXCR4 co-receptor func-tion. *J. Gen. Virol.* **82**, 1807–14.

26. Kazmierski WM, Kenakin TP, Gudmundsson KS, (2006). Peptide, peptidomimetic and small-molecule drug discovery targeting HIV-1 host-cell attachment and entry through gp120, gp41, CCR5 and CXCR4. *Chem. Biol. Drug Des.* **67**, 13–26.

27. Naldini L, Blomer U, Gallay P, Ory D, Mulligan R, Gage FH, Verma IM, Trono D (1996). In vivo gene delivery and stable trans-duction of nondividing cells by a lentiviral vector. *Science* **272**, 263–7.

28. Reiser J, Harmison G, Kluepfel-Stahl S, Brady RO, Karlsson S, Schubert M (1996). Transduction of nondividing cells using pseudotyped defective high-titer HIV type 1 particles. *Proc. Natl. Acad. Sci. U.S.A.* **93**, 15266–71.

29. Akkina RK, Walton RM, Chen ML, Li QX, Planelles V, Chen IS (1996). High-efficiency gene transfer into CD34+ cells with a human immunodeficiency virus type 1-based retrovi-ral vector pseudotyped with vesicular stomati-tis virus envelope glycoprotein G. *J. Virol.* **70**, 2581–5.

30. Steven AC, Spear PG (2006). Biochemistry. Viral glycoproteins and an evolutionary conundrum. *Science* **313**, 177–8.

31. Sharkey CM, North CL, Kuhn RJ, Sanders DA (2001). Ross River virus glycoprotein-pseudotyped retroviruses and stable cell lines for their production. *J. Virol.* **75**, 2653–9.

32. Picard-Maureau M, Jarmy G, Berg A, Rethwilm A, Lindemann D (2003). Foamy virus envelope glycoprotein-mediated entry involves a pH-dependent fusion process. *J. Virol.* **77**, 4722–30.

33. Bertrand P, Cote M, Zheng YM, Albritton LM, Liu SL (2008). Jaagsiekte sheep retrovi-rus utilizes a pH-dependent endocytosis path-way for entry. *J. Virol.* **82**, 2555–9.

34. Stein BS, Gowda SD, Lifson JD, Penhallow RC, Bensch KG, Engleman EG (1987). pH-independent HIV entry into CD4-positive T cells via virus envelope fusion to the plasma membrane. *Cell* **49**, 659–68.

35. Bartosch B, Vitelli A, Granier C, Goujon C, Dubuisson J, Pascale S, Scarselli E, Cortese R, Nicosia A, Cosset FL (2003). Cell entry of hepatitis C virus requires a set of co-receptors that include the CD81 tetraspanin and the SR-B1 scavenger receptor. *J. Biol. Chem.* **278**, 41624–30.

36. White LK, Yoon JJ, Lee JK, Sun A, Du Y, Fu H, Snyder JP, Plemper R (2007). Nonnucleoside inhibitor of measles virus RNA-dependent RNA polymerase complex activity. *Antimicrob. Agents Chemother.* **51**, 2293–303.

37. Burns JC, Friedmann T, Driever W, Burrascano M, Yee J-K (1993). Vesicular stomatitis

virus G glycoprotein pseudotyped retroviral vectors, concentration to very high titer and efficient gene transfer into mammalian and nonmammalian cells. *Proc. Natl. Acad. Sci. U.S.A.* **90**, 8033–37.

38. Takeuchi Y, Simpson G, Vile RG, Weiss RA, Collins MK (1992). Retroviral pseudotypes produced by rescue of a Moloney murine leukemia virus vector by C-type, but not D-type, retroviruses. *Virology* **186**, 792–4.

39. Lindemann D, Bock M, Schweizer M, Rethwilm A (1997). Efficient pseudotyping of murine leukemia virus particles with chimeric human foamy virus envelope proteins. *J. Virol.* **71**, 4815–20.

40. Mammano F, Salvatori F, Indraccolo S, De Rossi A, Chieco-Bianchi L, Gottlinger HG (1997). Truncation of the human immunodeficiency virus type 1 envelope glycoprotein allows efficient pseudotyping of Moloney murine leukemia virus particles and gene transfer into CD4+ cells. *J. Virol.* **71**, 3341–45.

41. Stitz J, Buchholz CJ, Engelstadter M, Uckert W, Bleimer U, Schmitt I, Cichutek K (2000). Lentiviral vectors pseudotyped with envelope glycoproteins derived from Gibbonr Ape Leukemia Virus and Murine Leukemia Virus 10A1. *Virology* **273**, 16–20.

42. Sandrin V, Boson B, Salmon P, Gay W, Negre D, Le Grand R, Trono D, Cosset F (2002). Lentiviral vectors pseudotyped with a modified RD114 envelope glycoprotein show increased stability in sera and augmented transduction of primary lymphocytes and CD34+ cells derived from human and nonhuman primates. *Blood* **100**, 823–32.

43. Di Nunzio F, Piovani B, Cosset F-L, Malivio F, Stornaiuolo A (2007). Transduction of human hematopoietic stem cells by lentiviral vectors pseudotyped with the RD114-TR chimeric envelope glycoprotein. *Hum. Gene Ther.* **18**, 811–20.

44. Hanawa H, Kelly PF, Nathwani AC, Persons DA, Vandergriff J, Hargrove P, Vanin EF, Nienhuis AW (2002). Comparison of various envelope proteins for their ability to pseudotype lentiviral vectors and transduce primitive hemaopoietic cells from human blood. *Mol. Ther.* **5**, 242–51.

45. Relander T, Johansson M, Olsson K, Ikeda Y, Takeuchi Y, Collins M, Richter J (2002). Gene transfer to repopulating human CD34+ cells using amphotropic-, GALV-, RD114-pseudotyped HIV-1 based vectors from stable producer cells. *Mol. Ther.* **11**, 452–9.

46. Sharkey CM, North CL, Kuhn RJ, Sanders DA (2001). Ross River virus glycoprotein-pseudotyped retroviruses and stable cell lines for their production. *J. Virol.* **75**, 2653–59.

47. Kang Y, Stein CS, Heth JA, Sinn PL, Penisten AK, Staber PD, Ratliff KL, Shen H, Barker CK, Martins I, et al. (2002). In vivo gene transfer using a nonprimate lentiviral vector pseudotyped with Ross River virus glycoproteins. *J. Virol.* **76**, 9378–88.

48. Kahl CA, Marsh J, Fyffe J, Sanders DA, Cornetta K (2004). Human immunodeficiency virus type 1-derived lentivirus vectors pseudotyped with envelope glycoproteins derived from Ross River virus and Semliki Forest virus. *J. Virol.* **78**, 1421–30.

49. Kang Y, S. SC, Heth JA, Sinn PL, Penisten AK, Stabler PD, Ratliff KL, Shen H, Barker CK, Martins I, et al. (2002). In vivo gene transfer using a nonprimate lentiviral vector pseudotyped with Ross River Virus glycoproteins. *J. Virol.* **76**, 9378–88.

50. Kahl CA, Pollok K, Haneline LS, Cornetta K (2005). Lentiviral vectors pseudotyped with glycoproteins from Ross River and vesicular stomatitis viruses, variable transduction related to cell type and culture conditions. *Mol. Ther.* **11**, 470–82.

51. Funke S, Maisner A, Muhlebach MD, Koehl U, Grez M, Cattaneo R, Cichutek K, Buchholz CJ (2008). Targeted cell entry of lentiviral vectors. *Mol. Ther.* **16**, 1427–36.

52. Sinn PL, Penisten AK, Burnight ER, Hickey MA, Williams G, McCoy DM, Mallampalli RK, McCray Jr PB (2005). Gene transfer to respiratory epithelia with lentivirus pseudotyped with Jaagsickte Sheep Retrovirus envelope glycoprotein. *Hum. Gene Ther.* **16**, 479–88.

53. Dull T, Zufferey R, Kelly M, Mandel RJ, Nguyen M, Trono D, Naldini L (1998). A third generation lentivirus with a conditional packaging system. *J. Virol.* **72**, 8463–71.

54. Kahl CA, Marsh J, Fyffe J, Sanders DA, Cornetta K (2004). Human immunodeficiency virus type 1-derived lentivirus vectors pseudotyped with envelope glycoproteins derived from Ross River virus and Semliki Forest virus. *J. Virol.* **78**, 1421–30.

55. Mochizuki H, Schwartz JP, Tanaka K, Brady RO, Reiser J (1998). High-titer human immunodeficiency virus type 1-based vector systems for gene delivery into nondividing cells. *J. Virol.* **72**, 8873–83.

56. Desmaris N, Bosch A, Salaun C, Petit C, Prevost MC, Tordo N, Perrin P, Schwartz O, de Rocquigny H, Heard JM (2001). Production and neurotropism of lentivirus vectors pseudotyped with lyssavirus envelope glycoproteins. *Mol. Ther.* **4**, 149–56.

57. Hanawa H, Kelly PF, Nathwani AC, Persons DA, Vandergriff JA, Hargrove P, Vanin EF, Nienhuis AW (2002). Comparison of various envelope proteins for their ability to pseudotype

lentiviral vectors and transduce primitive hematopoietic cells from human blood. *Mol. Ther.* **5**, 242–51.

58. Watson DJ, Kobinger GP, Passini MA, Wilson JM, Wolfe JH, (2002). Targeted transduction patterns in the mouse brain by lentivirus vectors pseudotyped with VSV, Ebola, Mokola, LCMV, or MuLV envelope proteins. *Mol. Ther.* **5**, 528–37.

59. Poeschla E, Gilbert J, Li X, Huang S, Ho A, Wong-Staal F (1998). Identification of a human immunodeficiency virus type 2 (HIV-2) encapsidation determinant and transduction of nondividing human cells by HIV-2-based lentivirus vectors. *J. Virol.* **72**, 6527–36.

60. Duisit G, Conrath H, Saleun S, Folliot S, Provost N, Cosset FL, Sandrin V, Moullier P, Rolling F (2002). Five recombinant simian immunodeficiency virus pseudotypes lead to exclusive transduction of retinal pigmented epithelium in rat. *Mol. Ther.* **6**, 446–54.

61. Mitrophanous K, Yoon S, Rohll J, Patil D, Wilkes F, Kim V, Kingsman S, Kingsman A, Mazarakis N (1999). Stable gene transfer to the nervous system using a non-primate lentiviral vector. *Gene Ther.* **6**, 1808–18.

62. Wong LF, Azzouz M, Walmsley LE, Askham Z, Wilkes FJ, Mitrophanous KA, Kingsman SM, Mazarakis ND (2004). Transduction patterns of pseudotyped lentiviral vectors in the nervous system. *Mol. Ther.* **9**, 101–11.

63. Olsen JC (1998). Gene transfer vectors derived from equine infectious anemia virus. *Gene Ther.* **5**, 1481–87.

64. McKay T, Patel M, Pickles RJ, Johnson LG, Olsen JC (2006). Influenza M2 envelope protein augments avian influenza hemagglutinin pseudotyping of lentiviral vectors. *Gene Ther.* **13**, 715–24.

65. Sandrin V, Boson B, Salmon P, Gay W, Negre D, Le Grand R, Trono D, Cosset FL (2002). Lentiviral vectors pseudotyped with a modified RD114 envelope glycoprotein show increased stability in sera and augmented transduction of primary lymphocytes and CD34+ cells derived from human and nonhuman primates. *Blood* **100**, 823–32.

66. Strang BL, Takeuchi Y, Relander T, Richter J, Bailey R, Sanders DA, Collins MK, Ikeda Y (2005). Human immunodeficiency virus type 1 vectors with alphavirus envelope glycoproteins produced from stable packaging cells. *J. Virol.* **79**, 1765–71.

67. Kang Y, Stein CS, Heth JA, Sinn PL, Penisten AK, Staber PD, Ratliff KL, Shen H, Barker CK, Martins I, et al. (2002). In vivo gene transfer using a nonprimate lentiviral vector pseudotyped with Ross River Virus glycoproteins. *J. Virol.* **76**, 9378–88.

68. Kolokoltsov AA, Weaver SC, Davey RA (2005). Efficient functional pseudotyping of oncoretroviral and lentiviral vectors by Venezuelan equine encephalitis virus envelope proteins. *J. Virol.* **79**, 756–63.

69. Poluri A, Ainsworth R, Weaver SC, Sutton RE (2008). Functional Pseudotyping of Human Immunodeficiency Virus Type 1 Vectors by Western Equine Encephalitis Virus Envelope Glycoprotein. *J. Virol.* **82**, 12580–4.

70. Morizono K, Bristol G, Xie YM, Kung SK, Chen IS (2001). Antibody-directed targeting of retroviral vectors via cell surface antigens. *J. Virol.* **75**, 8016–20.

71. Sinn PL, Hickey MA, Staber PD, Dylla DE, Jeffers SA, Davidson BL, Sanders DA, McCray Jr PB (2003). Lentivirus vectors pseudotyped with filoviral envelope glycoproteins transduce airway epithelia from the apical surface independently of folate receptor alpha. *J. Virol.* **77**, 5902–10.

72. Strang BL, Ikeda Y, Cosset FL, Collins MK, Takeuchi Y (2004). Characterization of HIV-1 vectors with gammaretrovirus envelope glycoproteins produced from stable packaging cells. *Gene Ther.* **11**, 591–8.

73. Christodoulopoulos I, Cannon PM (2001). Sequences in the cytoplasmic tail of the gibbon ape leukemia virus envelope protein that prevent its incorporation into lentivirus vectors. *J. Virol.* **75**, 4129–38.

74. Di Nunzio F, Piovani B, Cosset FL, Mavilio F, Stornaiuolo A (2007). Transduction of human hematopoietic stem cells by lentiviral vectors pseudotyped with the RD114-TR chimeric envelope glycoprotein. *Hum. Gene Ther.* **18**, 811–20.

75. Zhang XY, La Russa VF, Reiser J (2004). Transduction of bone-marrow-derived mesenchymal stem cells by using lentivirus vectors pseudotyped with modified RD114 envelope glycoproteins. *J. Virol.* **78**, 1219–29.

76. Schambach A, Galla M, Modlich U, Will E, Chandra S, Reeves L, Colbert M, Williams DA, von Kalle C, Baum C (2006). Lentiviral vectors pseudotyped with murine ecotropic envelope, increased biosafety and convenience in preclinical research. *Exp. Hematol.* **34**, 588–92.

77. Landau NR, Page KA, Littman DR (1991). Pseudotyping with human T-cell leukemia virus type I broadens the human immunodeficiency virus host range. *J. Virol.* **65**, 162, 9.

78. Liu SL, Halbert CL, Miller AD (2004). Jaagsiekte sheep retrovirus envelope efficiently pseudotypes human immunodeficiency virus

type 1-based lentiviral vectors. *J. Virol.* **78**, 2642–47.

79. Lewis BC, Chinnasamy N, Morgan RA, Varmus HE (2001). Development of an avian leukosis-sarcoma virus subgroup A pseudotyped lentiviral vector. *J. Virol.* **75**, 9339–44.

80. Frecha C, Costa C, Negre D, Gauthier E, Russell SJ, Cosset FL, Verhoeyen E (2008). Stable transduction of quiescent T-cells without induction of cycle progression by a novel lentiviral vector pseudotyped with measles virus glycoproteins. *Blood* **112**, 4843–52.

81. Jung C, Grzybowski BN, Tong S, Cheng L, Compans RW, Le Doux JM (2004). Lentiviral vectors pseudotyped with envelope glycoproteins derived from human parainfluenza virus type 3. *Biotechnol. Prog.* **20**, 1810–16.

82. Kowolik CM, Yee JK (2002). Preferential transduction of human hepatocytes with lentiviral vectors pseudotyped by Sendai virus F protein. *Mol. Ther.* **5**, 762–9.

83. Kobayashi M, Iida A, Ueda Y, Hasegawa M (2003). Pseudotyped lentivirus vectors derived from simian immunodeficiency virus SIVagm with envelope glycoproteins from paramyxovirus. *J. Virol.* **77**, 2607–14.

84. Kobinger GP, Deng S, Louboutin JP, Vatamaniuk M ,Matschinsky F, Markmann JF, Raper SE, Wilson JM (2004). Transduction of human islets with pseudotyped lentiviral vectors. *Hum. Gene Ther.* **15**, 211–9.

85. Hsu M, Zhang J, Flint M, Logvinoff C, Cheng-Mayer C, Rice CM, McKeating JA (2003). Hepatitis C virus glycoproteins mediate pH-dependent cell entry of pseudotyped retroviral particles. *Proc. Natl. Acad. Sci. U.S.A.* **100**, 7271–6.

86. Kumar M, Bradow BP, Zimmerberg J (2003). Large-scale production of pseudotyped lentiviral vectors using baculovirus GP64. *Hum. Gene Ther.* **14**, 67–77.

87. Kobinger GP, Weiner DJ, Yu QC, Wilson JM (2001). Filovirus-pseudotyped lentiviral vector can efficiently and stably transduce airway epithelia in vivo. *Nat. Biotechnol.* **19**, 225–30.

88. Azzouz M, Le T, Ralph GS, Walmsley L, Monani UR, Lee DC, Wilkes F, Mitrophanous KA, Kingsman SM, Burghes AH, Mazarakis ND (2004). Lentivector-mediated SMN replacement in a mouse model of spinal muscular atrophy. *J. Clin. Invest.* **114**, 1726–31.

89. Azzouz M, Ralph GS, Storkebaum E, Walmsley LE, Mitrophanous KA, Kingsman SM, Carmeliet P, Mazarakis ND (2004). VEGF delivery with retrogradely transported lentivector prolongs survival in a mouse ALS model. *Nature* **429**, 341–7.

90. Blomer U, Naldini L, Kafri T, Trono D, Verma IM, Gage FH (1997). Highly efficient and sustained gene transfer in adult neurons with a lentivirus vector. *J. Virol.* **71**, 6641–49.

91. Stein CS, Martins I, Davidson BL (2005). The lymphocytic choriomeningitis virus envelope glycoprotein targets lentiviral gene transfer vector to neural progenitors in the murine brain. *Mol. Ther.* **11**, 382–9.

92. Miletic H, Fischer YH, Neumann H, Hans V, Stenzel W, Giroglou T, Hermann M, Deckert M, Von Laer D (2004). Selective transduction of malignant glioma by lentiviral vectors pseudotyped with lymphocytic choriomeningitis virus glycoproteins. *Hum. Gene Ther* **15**, 1091–100.

93. Relander T, Johansson M, Olsson K, Ikeda Y, Takeuchi Y, Collins M, Richter J (2004). Gene transfer to repopulating human CD34+ cells using amphotropic-, GALV-, or RD114-pseudotyped HIV-1-based vectors from stable producer cells. *Mol. Ther.* **11**, 452–9.

94. MacKenzie TC, Kobinger GP, Kootstra NA, Radu A, Sena-Esteves M, Bouchard S, Wilson JM, Verma IM, Flake AW (2002). Efficient transduction of liver and muscle after in utero injection of lentiviral vectors with different pseudotypes. *Mol. Ther.* **6**, 349–58.

95. Auricchio A, Kobinger G, Anand V, Hildinger M, O'Connor E, Maguire AM, Wilson JM, Bennett J (2001). Exchange of surface proteins impacts on viral vector cellular specificity and transduction characteristics, the retina as a model. *Hum. Mol. Genet.* **10**, 3075–81.

96. Park F (2003). Correction of bleeding diathesis without liver toxicity using arenaviral-pseudotyped HIV-1-based vectors in hemophilia A mice. *Hum. Gene Ther.* **14**, 1489–94.

Chapter 4

Doxycycline Regulated Lentiviral Vectors

David Markusic and Jurgen Seppen

Abstract

Lentiviral vectors are a powerful tool to achieve regulated expression of transgenes in vivo and in vitro. The doxycycline-inducible system is well characterized and can be used to regulate expression mediated by lentiviral vectors. Because many different doxycycline-inducible lentiviral vectors have been described, choosing the best vector system can be difficult. This chapter can be used as a guide to select the optimal system for a particular application.

Key words: Doxycycline, Inducible Expression, Regulated Expression, Double cell transduction, RNA interference

1. Introduction

Selected applications of lentiviral gene transfer require the ability to regulate expression. In basic research, temporal expression of a gene can be needed or transgene toxicity could necessitate the ability to turn on and off expression.

Clinical implementation of gene therapy could also require the ability to modulate expression of the therapeutic gene to maintain expression levels within a therapeutic window and adjust expression levels based on disease progression within the patient (1).

To accomplish regulated gene expression, a variety of transcriptional regulatory systems were developed that can induce gene expression in response to an exogenously administered small molecule. The tetracycline-dependent transcriptional regulatory system (2) is one of the best-studied systems with proven efficacy in vitro and in vivo (1, 3–6). This system is based on the *Escherichia coli* Tn10 Tetracycline resistance operator consisting of the tetracycline repressor protein (TetR) and a specific DNA-binding site,

Maurizio Federico (ed.), *Lentivirus Gene Engineering Protocols,* Methods in Molecular Biology, vol. 614,
DOI 10.1007/978-1-60761-533-0_4, © Humana Press, a part of Springer Science+Business Media, LLC 2010

the tetracycline operator sequence (TetO). In the absence of tetracycline, the TetR protein dimerizes and binds to the TetO, thereby preventing gene expression. Tetracycline or doxycycline (i.e., a tetracycline derivative) can bind and induce a conformational change in the TetR protein leading to its disassociation from the TetO and activation of expression. A TetR mutant (rTetR) was identified with a reverse phenotype where binding to the TetO requires the presence of doxycycline. In the absence of doxycycline, this TetR mutant does not bind to the TetO (7).

To adapt this system for use in mammalian cells, the VP16 activation domain from Herpes simplex virus was fused to TetR (tTA) or the mutant TetR (rtTA) (2, 7). A tetracycline-responsive promoter (TRE) for mammalian expression was constructed by fusing a minimal cytomegalovirus (CMV) promoter to seven tandem TetO repeats (2, 7).

Thus, in the tTA system, expression from the TRE is turned off by adding doxycycline (Tet-off, Fig. 1) and in the rtTA system expression is turned on by adding doxycycline (Tet-on, Fig. 1). However, the original mutant rtTA was not very sensitive to doxycycline. Mutagenesis and codon optimization of the reverse TetR produced improved rtTA variants with reduced basal activity and improved sensitivity to doxycycline (8). Additional rtTA variants were identified by viral evolution that can respond to as little as 10 ng/ml doxycycline (9). These new variants now have a similar sensitivity to doxycycline as the original tTA.

As the TRE has some basal activity, a further improvement was the inclusion of a repressor of expression in the system.

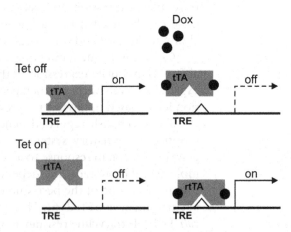

Fig. 1. Schematic representation of the Tet-off and Tet-on doxycycline-inducible systems. In the Tet-off system (upper panel), the tTA protein only binds to the tetracycline responsive promoter (*TRE*) in the *absence* of doxycycline (*Dox.*) and activates transcription. When doxycycline is present, tTA dissociates from the TRE and transcription is halted. In the Tet-on system (*lower panel*), the rtTA protein only binds to the TRE in the presence of doxycycline and activates transcription. When doxycycline is withdrawn, rtTA dissociates from the TRE and transcription is halted

Kruppel-associated box (krab) domains can locally repress transcription when bound to DNA. By fusing TetR to a krab domain from the kidney protein kid-1, a doxycycline responsive repressor (tTS) was obtained that can repress expression from TRE (10). The tTS was generated with the regular and reverse TetR, thus generating rtTS and tTS (11). The doxycycline-dependent repressors can be used in combination with the reverse versions of tTA and rtTA, but can also be used as the only regulatory molecule.

2. Doxycycline-Regulated Lentiviral Vectors

Earlier described doxycycline-regulated lentiviral vectors used two separate constructs, one containing the TRE-regulated transgene and the second containing rtTA expressed by a constitutive promoter (Fig. 1, panels 1a and 1b). This binary system relies on the co-transduction of cells with two different vectors. Cells can be transduced first with a vector mediating constitutive expression of tTA or rtTA and subsequently with a vector containing a transgene under control of the TRE. Selection and screening can be used to obtain a homogenously transduced population displaying the desired induction kinetics. High levels of rTA and rtTA expression can be toxic, likely as the consequence of the VP16 domain. The binary approach, thus, allows for the selection of cells expressing sufficient, but nontoxic amounts of the transactivator.

However, simultaneous or serial lentiviral transductions are not possible in vivo or in difficult to transduce cells. Thus, lentiviral vectors were developed that incorporate all elements of the doxycycline-inducible system (Fig. 2, panels 2 and 3).

A disadvantage of a single vector is promoter interference occurring between the TRE promoter and the constitutive promoter used to express tTA or rtTA. This can lead to increased background levels. To solve this problem, tTA or rtTA can be expressed from the TRE promoter resulting in autoregulatory expression (Fig. 2, panel 3) (12).

3. Selection of the Optimal Doxycycline-Inducible Lentiviral Vector

A large number of different doxycycline-inducible lentiviral vectors are available. Fortunately, selection of the optimal system can partially be inferred from the experimental setting. Below, we summarize the advantages and disadvantages of different doxycycline-inducible lentiviral vectors. This section can act as a guide in choosing the optimal system for an individual application.

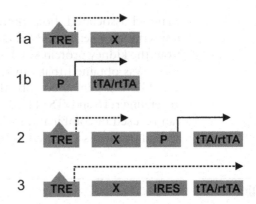

Fig. 2. Schematic representation of binary and single doxycycline-inducible lentiviral vectors. In a binary inducible vector system (*panels 1a, b*), one vector contains the TRE, controlling expression of transgene X, (*panel 1a*). The other vector contains tTA or rtTA under control of a constitutive promoter P (*panel 1b*). Regulated expression requires the transduction of target cells with both vectors. First-generation single-vector systems (*panel 2*) contain both expression cassettes of the binary system in one vector backbone. In advanced single-vector systems (*panel 3*), the transgene and tTA/rtTA are under control of the same TRE, linked by an internal ribosome entry site (IRES)

In general, it is advisable to use the most simple system that is meeting the need of the particular application. Complicated single vectors can be necessary, but the inclusion of multiple elements in the vector increases the difficulty of cloning the desired construct and decreases the available space for transgenes and marker genes (i.e., GFP) or drug selection genes. More complicated vector designs and larger expression cassettes also have a tendency to generate low titer vector preparations.

4. Tet-on or Tet-off

In most applications, it is preferable to minimize the duration of doxycycline exposure. Therefore, the choice between Tet-on and Tet-off depends on whether the system will be studied primarily in the repressed or induced state.

In the past, tTA was often preferred because lower concentrations of doxycycline were required. However, new variants of rtTA approach the doxycycline sensitivity of tTA, and comparable concentrations of doxycycline can be used with both Tet on and Tet off systems. Although the concentration of doxycycline needed to induce or repress expression in vitro and in vivo are usually well tolerated, it will depend on the application if cells or animals can be maintained for longer periods under doxycycline pressure.

5. Regulated Expression In vitro

When target cells are readily transduced by lentiviral vectors and can be expanded and cloned before use, a two-vector system is preferable (Fig. 2, panels 1a and 1b). High levels of tTA or rtTA expression can be toxic. After transduction with a tTA or rtTA vector, cells can be selected for sufficient but non toxic expression levels of the regulatory proteins. In a one-vector system, the induction of the therapeutic protein will also lead to induction of tTA or rtTA because the VP16 domain will also have a local inducing effect on the promoter driving expression of the regulatory protein (Fig. 2, panels 2 and 3). With a two vector system, the regulatory protein expression cassette is spatially separated from the inducible protein and this effect cannot occur.

Background expression levels of the TRE will vary with chromosomal location. Thus, an advantage of a two-vector system is that clones can be isolated that have low background activity. An additional advantage is that double-vector systems also allow the use of larger genes because more space is available in the vector.

If the target cell type is difficult to be transduced or when primary cells that cannot be expanded are studied, one-vector systems should be considered for in vitro use.

6. Regulated Expression In vivo

The simultaneous transduction of a single cell with two different vectors is likely difficult in vivo. Therefore, for direct injection into animals, a single vector system is preferred.

When cells can be transduced ex vivo and are subsequently transplanted, a two-vector system can be used. However, this should only be considered when the target cells are efficiently transduced and multiple integrations per cell can be easily obtained.

7. Minimizing Background Expression

Because the TRE promoter is slightly leaky, the choice of vector also depends on the necessity to completely abolish background expression. Using advanced single vectors or by careful selection of clones in a two-vector system, background expression can be reduced, but not completely abolished. Therefore, when it is essential to completely shut off basal expression, it is necessary to use vectors that include a TetO-binding repressor protein, such as tTS or rtTS (11).

8. Doxycycline-Regulated RNA Interference

RNA interference is a powerful technique to silence gene expression and lentiviral vectors are excellently suited to deliver RNA interference cassettes.

Several methods have been developed to achieve doxycycline-inducible gene silencing. When short hairpin siRNAs are generated by the use of polymerase III promoters, doxycycline regulation can be achieved by two different mechanisms. By placing the TetO sequence in the pol III promoter, a binding site for tTA, rtTA, tTS, or rtTS is generated. Using tTA or rtTA, transcription from the pol III promoter is inhibited by steric hindrance after binding of the transactivator. Alternatively, tTS or rtTS binding can be used to silence expression of the pol III promoter-driven siRNA (13). Both approaches have been incorporated in single- and double-vector systems.

A different and more effective approach is to use microRNA sequences that have been adapted to incorporate siRNA sequences that silence genes of interest (14). This approach has the advantage that the modified microRNAs can be generated from a polymerase II promoter and thus an inducible vector with TRE promoter can be used for doxycycline-inducible silencing. In addition, multiple-siRNA-sequences can be incorporated in a single transcript. However, cloning of microRNA based RNA interference cassettes is more time-consuming than the use of short hairpin microRNAs.

9. Administration of Doxycycline in Cell Culture and in Animals

To prepare stock solutions, dissolve doxycycline-hyclate in deionized water at a concentration range of 1–50 mg/ml. Concentration depends on the intended route of administration and final concentration needed. Solutions should be filter-sterilized and can be kept at 4°C protected from light for up to 4 weeks. For long-term storage, doxycycline stock solutions should be kept at −20°C. It is not recommended to dissolve doxycycline for stock solutions in PBS, as this may lead to precipitation. Fresh doxycycline working solutions are recommended, as long-term storage of diluted doxycycline may also lead to precipitation. For intraperitoneal (IP) injections, doxycycline freshly diluted into PBS can be used.

For the administration in cell culture, doxycycline can be directly added to cell cultures from sterile-filtered stock solutions. The optimal concentration of doxycycline has to be determined empirically as this depends on the type of transactivator used and the target cell line. Both tTA and advanced versions of tTA should be activated by concentrations of 1 μg/ml. However, doxycycline is toxic to some cell types and concentrations of as low as 10 ng/ml can also give a good response.

The doxycycline-inducible system has been shown to work well in many different in vivo systems. However, the chimeric bacterial and viral transactivators can be a potent immunogen. Indeed, in studies performed in mice and nonhuman primates (15, 16), potent immune responses were detected against rtTA, including rtTA antibodies and potent cytotoxic T-cell responses, leading to immune-mediated clearance of transduced cells and loss of transgene expression. This immunogenicity must be kept in mind when designing experiments with doxycycline-regulated lentiviral vectors in immune-competent animals that are not tolerized to tTA or rtTA.

Doxycycline can be administered to animals by different routes, i.e., IP injections, in the drinking water, or in food pellets. In the case doxycycline is provided in drinking water and in food *ad libitum*, this results in little control over dosing of doxycycline. A direct method of administration, such as IP injections (10–50 mg/kg), gives much more control over doxycycline concentrations in vivo. The exact amount of administered doxycycline should be empirically determined for toxicity in the experimental animal model used and for the required levels of doxycycline in the target tissue. However, because daily injections are needed, IP administration is the most labor-intensive approach and may not be suitable for long-term studies.

For the administration in drinking water, the doxycycline stock solution should be diluted in drinking water to a concentration of 200 mg/l. Water bottles should be protected from light and freshly prepared doxycycline water should be given every 3–4 days. The final concentration of doxycycline in water will depend on the selection of transactivator and tissue penetration of doxycycline in the transduced target tissue.

Some studies indicated the need to include 5% sucrose in the drinking water to counteract the bitter taste of doxycycline. However, our experience in rats and a recent study in mice (17) suggest that the addition of 5% sucrose leads to overconsumption of water, which can have consequences on the intake levels of doxycycline, leading to both disruption in normal food consumption, and excessive urination. Indeed, in mice, it was shown that the absence of sucrose resulted in nominal water intake. Thus, we caution the use of sucrose in doxycycline water and recommend to initially prepare doxycycline water without sucrose. If water consumption is below nominal intake levels, titrate up the amount of supplemented sucrose necessary to restore nominal intake levels.

Finally, for the administration in food, specially formulated rodent food pellets containing doxycycline at 200 mg/kg or other concentrations are commercially available. The appropriate concentration of doxycycline in food will depend on the selection of transactivator and tissue penetration of doxycycline in the transduced target tissue. However, doses of doxycycline up to 2 g/kg do not cause problems in mice. Administration of doxycycline in food requires the least amount of effort, but it is difficult to control doxycycline levels. Furthermore, the doxycycline containing food might not have the same composition as the standard chow of the animal facility. In that case, the corresponding doxycycline free food has to be given to the control animals as well.

References

1. Clackson T. (2000) Regulated gene expression systems. *Gene Ther* 7, 120–5.

2. Gossen M, and Bujard H. (1992) Tight control of gene expression in mammalian cells by tetracycline-responsive promoters. *Proc Natl Acad Sci U S A* 89, 5547–51.

3. Agha-Mohammadi S, and Lotze MT. (2000) Regulatable systems: applications in gene therapy and replicating viruses. *J Clin Invest* 105, 1177–83.

4. Corbel SY, and Rossi FM. (2002) Latest developments and in vivo use of the Tet system: ex vivo and in vivo delivery of tetracycline-regulated genes. *Curr Opin Biotechnol* 13, 448–52.

5. Goverdhana S, Puntel M, Xiong W et al. (2005) Regulatable gene expression systems for gene therapy applications: progress and future challenges. *Mol Ther* 12, 189–211.

6. Toniatti C, Bujard H, Cortese R, and Ciliberto G. (2004) Gene therapy progress and prospects: transcription regulatory systems. *Gene Ther* 11, 649–57.

7. Gossen M, Freundlieb S, Bender G, Muller G, Hillen W, and Bujard H. (1995) Transcriptional activation by tetracyclines in mammalian cells. *Science* 268, 1766–69.

8. Urlinger S, Baron U, Thellmann M, Hasan MT, Bujard H, and Hillen W. (2000) Exploring the sequence space for tetracycline-dependent transcriptional activators: novel mutations yield expanded range and sensitivity. *Proc Natl Acad Sci U S A* 97, 7963–8.

9. Zhou X, Vink M, Klaver B, Berkhout B, and Das AT. (2006) Optimization of the Tet-On system for regulated gene expression through viral evolution. *Gene Ther* 13, 1382–90.

10. Freundlieb S, Schirra-Muller C, and Bujard H. (1999). A tetracycline controlled activation/repression system with increased potential for gene transfer into mammalian cells. *J Gene Med* 1, 4–12.

11. Szulc J, Wiznerowicz M, Sauvain MO, Trono D, and Aebischer P. (2006). A versatile tool for conditional gene expression and knockdown. *Nat Methods* 3, 109–16.

12. Markusic D, Oude-Elferink R, Das AT, Berkhout B, and Seppen J. (2005) Comparison of single regulated lentiviral vectors with rtTA expression driven by an autoregulatory loop or a constitutive promoter. *Nucleic Acids Res* 33, e63.

13. Wiznerowicz M, Szulc J, and Trono D. (2006) Tuning silence: conditional systems for RNA interference. *Nat Methods* 3, 682–8.

14. Zeng Y, Wagner EJ, and Cullen BR. (2002) Both natural and designed micro RNAs can inhibit the expression of cognate mRNAs when expressed in human cells. *Mol Cell* 9, 1327–33.

15. Ginhoux F, Turbant S, Gross DA et al. (2004) HLA-A*0201-restricted cytolytic responses to the rtTA transactivator dominant and cryptic epitopes compromise transgene expression induced by the tetracycline on system. *Mol Ther* 10, 279–89.

16. Latta-Mahieu M, Rolland M, Caillet C et al. (2002) Gene transfer of a chimeric trans-activator is immunogenic and results in short-lived transgene expression. *Hum Gene Ther* 13, 1611–20.

17. Hojman P, Eriksen J, and Gehl J. (2007) Tet-On induction with doxycycline after gene transfer in mice: sweetening of drinking water is not a good idea. *Anim Biotechnol* 18, 183-8.

Chapter 5

Strategies to Insulate Lentiviral Vector-Expressed Transgenes

Ali Ramezani and Robert G. Hawley

Abstract

Lentiviruses are capable of infecting many cells irrespective of their cycling status, stably inserting DNA copies of the viral RNA genomes into host chromosomes. This property has led to the development of lentiviral vectors for high-efficiency gene transfer to a wide variety of cell types, from slowly proliferating hematopoietic stem cells to terminally differentiated neurons. Regardless of their advantage over gammaretroviral vectors, which can only introduce transgenes into target cells that are actively dividing, lentiviral vectors are still susceptible to chromosomal position effects that result in transgene silencing or variegated expression. In this chapter, various genetic regulatory elements are described that can be incorporated within lentiviral vector backbones to minimize the influences of neighboring chromatin on single-copy transgene expression. The modifications include utilization of strong internal enhancer–promoter sequences, addition of scaffold/matrix attachment regions, and flanking the transcriptional unit with chromatin domain insulators. Protocols are provided to evaluate the performance as well as the relative biosafety of lentiviral vectors containing these elements.

Key words: Lentiviral vectors, Position effects, Transgene silencing, Insulators, Enhancer-blocking elements, Scaffold/matrix attachment regions, Insertional mutagenesis, Genotoxicity

1. Introduction

Gammaretroviral vectors derived from Moloney murine leukemia virus (MLV) integrate into the host cell genome and therefore have been widely used for stable gene transfer (1). One of the main limitations of gammaretroviral vectors, however, is their dependence on nuclear membrane dissolution during mitosis for access to and integration into target cell chromosomes (2). By comparison, lentiviruses, such as human immunodeficiency virus type 1 (HIV-1), have the capacity to infect many nondividing

Maurizio Federico (ed.), *Lentivirus Gene Engineering Protocols*, Methods in Molecular Biology, vol. 614,
DOI 10.1007/978-1-60761-533-0_5, © Humana Press, a part of Springer Science+Business Media, LLC 2010

as well as dividing cells (3, 4). This property prompted the development of HIV-1-based lentiviral vectors that can efficiently transduce slowly proliferating cells such as hematopoietic stem cells as well as terminally differentiated cells such as neurons (1, 5, 6).

Another drawback of conventional MLV-based gammaretroviral vectors is their frequent silencing after transduction of certain cell types, particularly embryonic stem cells (ESCs) (7–9). Silencing elements have been identified in the MLV long terminal repeat (LTR) (10, 11) and the primer-binding site (12–14), the removal of which has resulted in improved gammaretroviral vectors that exhibit an increased probability of transgene expression in ESCs and other primitive cells (1, 15–18). Additionally, position effects due to chromatin structure at the vector insertion site contribute to gammaretroviral vector silencing shortly after integration, to expression variegation (in which variable expression is observed within the clonal progeny of a target cell), and to extinction of expression (in which expression is downregulated with time or during differentiation) (19–22). Accumulating data suggest that lentiviral vectors are also susceptible to stem cell-associated and chromosomal position effects that result in variegated expression and silencing (23–29).

Progress in the field of gene expression and chromatin structure has led to the identification of genetic regulatory elements capable of counteracting position effects (reviewed in ref. 1): strong enhancer-promoters allow maintenance of active chromatin configurations, even at unfavorable heterochromatin insertion sites; scaffold/matrix attachment regions (S/MARs) serve to organize chromatin into structural domains via anchorage to the nucleoskeleton; and chromatin domain insulators act as barrier elements that can physically block the influence of repressive chromatin structure. Most insulators also posses enhancer blocking activity distinct from their anti-silencer properties.

1.1. Strong Enhancer–Promoter Sequences

Silencing is facilitated by a large number of protein factors, which in some cases can be antagonized by proteins such as transcriptional activators (30). For example, some strong promoters have been shown to counteract chromosomal position effects or withstand silencing induced during differentiation (31, 32). Enhancers and promoters are believed to be foci for changes in chromatin structure, although the mechanisms underlying these changes are incompletely understood (33–35). In cultured cells, it has been shown that enhancers could retard silencing of integrated reporter genes without significantly affecting the level at which they are expressed (36, 37). These results suggest that enhancers function to disrupt or prevent the formation of repressive chromatin structures, thus permitting maintenance of gene expression.

Some of the commonly used strong promoters include the murine stem cell virus (MSCV) LTR (17), the human elongation factor 1α (EF1α) promoter (38, 39), and the composite promoter (CAG) composed of the cytomegalovirus (CMV) immediate early region enhancer linked to chicken β-actin promoter sequences (40). The MSCV LTR is permissive for expression in murine and human hematopoietic stem/progenitor cells (HSPCs) with long-term hematopoietic repopulating potential (18, 41–44). The EF1α promoter drives widespread expression in vivo (45) and has been identified as a strong promoter in primary human CD34+ HSPCs (46, 47). The composite CAG promoter functions efficiently in murine ESCs and in transgenic mice (48, 49). We previously developed a series of self-inactivating (SIN) HIV-1-based lentiviral vectors containing a variety of strong viral and cellular enhancer-promoters (43, 50). Of those evaluated, the MSCV LTR, the EF1α promoter, and the CAG promoter consistently mediated high and sustained transgene expression in the cell types studied.

We subsequently described an MSCV-based SIN gammaretroviral vector, MSinSB, which utilizes a modified version of the composite CAG promoter (32). When used to transduce murine ESCs, the MSinSB vector directed high transgene expression levels, which were maintained during in vitro hematopoietic differentiation of the cells. This was in sharp contrast to the parental MSCV gammaretroviral vector, which was tested in parallel and found to almost completely silence during the ESC differentiation process. Persistent high-level expression of the MSinSB gammaretroviral vector was also demonstrated in murine bone marrow transplant recipients and following in vitro myelomonocytic differentiation of human CD34+ cord blood HSPCs.

1.2. Scaffold/Matrix Attachment Regions (S/MARs)

Eukaryotic cells have evolved specialized DNA elements known as boundary elements that mark the borders of adjacent chromatin domains and protect active regions from the repressive effects of nearby heterochromatin. One class of element, termed S/MARs, anchors chromatin to a skeleton of protein cross-ties called the nuclear scaffold (in metaphase) or nuclear matrix (in interphase) (51–57). It has been suggested that S/MARs function by forming chromatin loops, bringing enhancers and other distal regulatory elements in close proximity to their corresponding promoter sequences. Additionally, they may augment transgene expression by protecting the enhancer from DNA methylation (58), which may involve localized attraction of histone acetylation (59). In this regard, insertion of the S/MAR believed to define the upstream border of the human interferon-β locus (IFN-SAR) into gammaretroviral vectors has been shown to result in improved transgene expression (55, 60–62).

We introduced the IFN-SAR into a series of HIV-1-based lentiviral vectors and analyzed reporter expression in transduced populations and individual clones of the human CD34+ HSPC line KG1a (50). After 4 months of continuous culture, significantly higher levels of transgene expression were observed from three different internal promoters in cells transduced with the modified vectors containing the IFN-SAR by comparison to the respective parental lentiviral vector backbones. Others have reported that inclusion of an S/MAR from the mouse immunoglobulin kappa light chain gene significantly increased the probability of transgene expression from SIN lentiviral vectors containing two different internal promoters (63).

1.3. Insulators

Chromatin domain insulators are DNA boundary elements capable of suppressing repressive position effects and in most cases also blocking the action of distant enhancers (64, 65). The best studied vertebrate element is a 1.2-kilobase (kb) fragment encompassing a DNase I hypersensitive site at the 5′ end (5′HS4) of the chicken β-globin locus (66, 67). Extensive characterization by Felsenfeld and colleagues has shown that the chicken β-globin 5′HS4 insulator can protect against chromosomal position effects and also provide enhancer blocking function (Fig. 1) (65–72). Almost all of the enhancer blocking activity of the 1.2-kb 5′HS4 insulator fragment can be conferred by a 42-bp sequence (DNase I-footprinted region II; FII) that is bound by the zinc finger protein CTCF (CCCTC-binding factor) (69, 71). Consistent with this role, analyzes of the genome-wide distribution of CTCF-binding sites demonstrate a tendency of CTCF to reside in loci that separate chromosomal domains (73), perhaps facilitating the formation of chromatin loops (72, 74). Recent studies have revealed that cohesin complexes, best known for sister chromatid cohesion, contribute to CTCF-dependent enhancer blocking (75, 76). It is important to point out, however, that the contribution of CTCF to the anti-silencer properties associated with the barrier function of insulators remains unclear (71, 76–78).

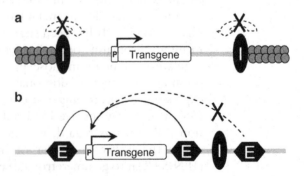

Fig. 1. Schematic diagram illustrating the two main and separate properties of insulators. (a) An insulator element (I) can protect against chromosomal position effects by blocking the advancement of adjacent repressive chromatin. (b) Many insulators also provide enhancer blocking function by preventing the activity of an enhancer (E), but only when placed between the enhancer and a promoter (P)

In an attempt to eliminate chromosomal position effects and transgene silencing, we and others have used insulators to flank gammaretroviral and lentiviral vectors. Some improvements were obtained when a single copy of the 1.2-kb fragment of the chicken β-globin 5'HS4 insulator was added (50, 77, 79–81). However, in our experience, the titers of 5'HS4 insulator-containing vectors were generally reduced (50). Moreover, Felsenfeld and colleagues reported superior protection in cell culture transfection and *Drosophila* transgenesis experiments when the transgene was flanked by two direct copies of the 1.2-kb 5'HS4 fragment (66). In further work by this group, the chicken β-globin 5'HS4 insulator was mapped to a 5' 250-bp fragment (encompassing the FII region containing the CTCF binding site), two copies of which were shown to provide complete insulator function (70). Therefore, as one approach toward alleviating ESC differentiation-associated gammaretroviral extinction, we inserted two copies of the 250-bp 5'HS4 core in tandem into the U3 region of the 3' LTR of an MSCV-based gammaretroviral vector (32). However, analysis of proviral DNA from stably transduced cells indicated that the direct repeat configuration was highly unstable, with one copy of the 5'HS4 insulator core being deleted at high frequency during replication. More recently, a 400-bp fragment consisting of the 5'HS4 insulator 250-bp core plus 150 bp of 3' flanking sequences has been reported to protect gammaretroviral vector and lentiviral vector transgene expression to the same degree as the full-length 1.2-kb fragment (82).

Given the ability of the chicken β-globin 5'HS4 insulator and the IFN-SAR to individually provide some protection against position effects to gammaretroviral vectors, and in view of their complementary properties and structural features – i.e., the core element of the 5'HS4 insulator is GC-rich with a high density of cytidine-guanosine (CpG) dinucleotides (reminiscent of a CpG island), whereas S/MARs are AT-rich sequences (57) – plus the fact that lentiviral vectors can readily accommodate large DNA inserts (83), we were interested in evaluating the utility of combining these elements (50). Our findings in the human CD34[+] HSPC line KG1a and in primary human CD34[+] cord blood cells demonstrated that position-effect protection of lentiviral vector-mediated transgene expression provided by combinatorial association of the 5'HS4 insulator and the IFN-SAR was superior to that conferred by either element alone. The lentiviral vector backbone utilizing the 5'HS4 insulator/IFN-SAR combination has also been successfully used to confer stable long-term transgene expression in human ESCs and their differentiated progeny (84, 85).

Kwaks et al. identified a number of novel human genetic elements that exhibit enhancer blocking activity in a screen of genomic DNA fragments that protected a reporter gene from repression by the chromatin-associated repressor proteins, heterochromatin protein 1, or the Polycomb group protein HPC2 (86). One of these so-called "anti-repressor" elements, element 40, has subsequently been combined with the IFN-SAR and shown to minimize variegated expression in lentiviral transgenic mice (87, 88).

Insulation of the transgene from the host cell genome has the reciprocal benefit of protecting the host cell genome against insertional mutagenesis caused by integrating vectors (i.e., reducing their genotoxic potential) (89–91). We recently reported the development of a 77-bp element, FII/BEAD-A (FB), which combines the minimal enhancer blocking components of the chicken β-globin 5′HS4 insulator (a modified 42-bp FII region containing the CTCF-binding site) and a homologous CTCF-binding site-containing region from the human T-cell receptor α/δ BEAD-1 insulator (92). With a new flow cytometry-based assay, we showed that the FB element was as effective in enhancer blocking activity as the prototypical 1.2-kb 5′HS4 insulator fragment. Notably, using an in vitro insertional mutagenesis assay involving primary murine HSPCs (93), we found that both SIN gammaretroviral and lentiviral vectors containing the FB element exhibited greatly reduced transforming potential – to background levels under the experimental conditions used – compared to their unshielded counterparts.

This chapter describes strategies to incorporate strong enhancer-promoter sequences, the IFN-SAR, and the chicken β-globin 5′HS4 insulator into lentiviral vectors, and provides protocols to evaluate insulator performance as well as the relative biosafety of the modified vectors.

2. Materials

2.1. Barrier Function Assay

2.1.1. Design and Construction of Reporter Plasmids for Barrier Function Assay

1. Plasmids: The reporter plasmid, pRL-TKGFP-CMVE, has been described previously (92). The plasmid and the corresponding plasmid map can be obtained from the authors upon request. Starting material plasmids: Herpes simplex virus thymidine kinase (TK) promoter plasmid (e.g., pRL-TK; Promega Corp., Madison, WI, USA); green fluorescent protein (GFP) gene plasmid (e.g., Clontech Laboratories, Inc., Mountain View, CA, USA, pAcGFP1-N1; Invitrogen Corp., Carlsbad, CA, USA, pcDNA™6.2/N-EmGFP-GW/TOPO®; Stratagene Corp., La Jolla, CA, USA, phrGFP II-N); CMV immediate early region enhancer plasmid (e.g., Clontech Laboratories, pCMV-DsRed-Express); chicken β-globin 5′HS4 insulator plasmid (e.g., pJC13-1 (66)).

2. Restriction and modifying enzymes.

3. DH5α competent bacteria (Invitrogen).

4. Luria-Bertani media (LB): LB + 50 μg/ml Ampicillin (LB-AMP); AMP-agar plates.

5. QIAquick Gel Extraction Kit; QIAGEN Plasmid Maxi Kit (Qiagen, Valencia, CA, USA).

2.1.2. Testing of Reporter Plasmids for Barrier Function

1. Cell line: K562 human erythroleukemia cells (American Type Culture Collection, Manassas, VA, USA, cat. no. CCL-243).

2. K562 culture medium: Iscove's modified Dulbecco's medium (IMDM; Mediatech, Herndon, VA, USA) plus 10% heat-inactivated fetal bovine serum (FBS; HyClone, Logan, Utah, USA), l-glutamine (2 mM), penicillin (50 IU/ml), and strep-tomycin (50 μg/ml).

3. Phosphate-buffered saline (PBS).

4. Electroporation cuvettes.

5. Electroporator.

6. Analytical flow cytometer (such as a FACSCalibur instru-ment, BD Biosciences, San Jose, CA, USA).

7. Fluorescence-activated cell sorter (such as a FACSAria instru-ment, BD Biosciences).

2.2. Enhancer Blocking Assay

2.2.1. Design and Construction of Reporter Plasmids for Enhancer Blocking Assay

1. Plasmids: The reporter plasmid, pRL-TKGFP-CMVE-TKRFP, has been described previously (92). The plasmid and the corresponding plasmid map can be obtained from the authors upon request. Starting material plasmids: pRL-TKGFP-CMVE and herpes simplex virus TK promoter plas-mid described in Subheading 2.1.1; red fluorescent protein (RFP) gene plasmid (e.g., pDsRedT4-N1 plasmid (94)).

2. Restriction and modifying enzymes.

3. DH5α competent bacteria (Invitrogen).

4. QIAquick Gel Extraction Kit, QIAGEN Plasmid Maxi Kit (Qiagen).

2.2.2. Testing of Reporter Plasmids for Enhancer Blocking Activity

1. Cell lines: 293T/17 (293T) human embryonic kidney cell line (American Type Culture Collection, cat. no. CRL-11268) (see Note 1); K562 cells (American Type Culture Collection, cat. no. CCL-243).

2. 293T culture medium: Dulbecco's Modified Eagle's Medium (DMEM; Mediatech) supplemented with 4.5 g/l glucose, 2 mM l-glutamine, 50 IU/ml penicillin, 50 μg/ml strepto-mycin, and 10% heat-inactivated FBS. Store at 4°C and warm up to 37°C before use.

3. K562 culture medium prepared as described in Sub-heading 2.1.2.

4. 2.5 M $CaCl_2$: Dissolve 183.7 g $CaCl_2$ dihydrate (tissue culture grade) in deionized, distilled water. Bring the volume up to 500 ml and filter-sterilize using a 0.22-μm nitrocellulose filter. Stable at –20°C.

5. 2× *N*-(2-hydroxyethyl)piperazine-*N′*-(2-ethanesulfonic acid) (HEPES)-buffered saline (2× HBS): 50 mM HEPES (Sigma-Aldrich Corp., St. Louis, MO, USA), 280 mM NaCl, 1.5 mM Na_2HPO_4. Titrate to pH 7.05 with 5 N NaOH. Filter-sterilize using a 0.22-μm nitrocellulose filter. Store as single use aliquots at –20°C.

6. PBS.

7. Electroporation cuvettes.

8. Electroporator.

9. Analytical flow cytometer.

2.3. Lentiviral Vector Construction

2.3.1. Strong Enhancer–Promoter Sequences

1. Plasmids: Lentiviral vector plasmid containing reporter gene, e.g., GFP (see Note 2); MSCV LTR promoter plasmid (17); human EF1α promoter plasmid (e.g., pEF-BOS (39)); CAG promoter plasmid (e.g., pBacMam-2; Novagen, Inc., Madison, WI, USA).

2. Restriction and modifying enzymes.

3. DH5α competent bacteria (Invitrogen).

4. QIAquick Gel Extraction Kit; QIAGEN Plasmid Maxi Kit (Qiagen).

2.3.2. S/MAR Element

1. Plasmids: Lentiviral vector plasmid containing reporter gene, e.g., GFP (see Note 2); IFN-SAR plasmid (e.g., pCL, (55)).

2. Restriction and modifying enzymes.

3. DH5α competent bacteria (Invitrogen).

4. QIAquick Gel Extraction Kit; QIAGEN Plasmid Maxi Kit (Qiagen).

2.3.3. 5′HS4 Insulator

1. Plasmids: Lentiviral vector plasmid containing reporter gene, e.g., GFP (see Note 2); chicken (-globin 5′HS4 insulator plasmid, e.g., pJC13-1, (66)).

2. Restriction and modifying enzymes.

3. DH5α competent bacteria (Invitrogen).

4. QIAquick Gel Extraction Kit, QIAGEN Plasmid Maxi Kit (Qiagen).

2.4. Lentiviral Vector Function

2.4.1. Transgene Expression in Cell Lines

1. Replication-defective lentiviral vector particles (95): produce and titer as described in other Chapters of this book.

2. Cell line: K562 cells.

3. K562 culture medium prepared as described in Subheading 2.1.2.

4. Polybrene stock solution (Sigma-Aldrich Corp.; hexadime-thrine bromide, cat. no. H9268). Prepare stock of 6 mg/ml (1000×) in sterile deionized, distilled water; aliquot and store at −20°C.

5. Analytical flow cytometer.

2.4.2. Stability of the Inserted Elements

1. Genomic DNA Extraction Kit (Sigma-Aldrich Corp.).

2. PCR primers (Sigma-Aldrich Corp.), 0.05 μmole scale.

3. Taq DNA polymerase, dNTP mix (Promega Corp.).

4. PCR thermocycler.

2.4.3. Transgene Expression Following Differentiation of Transduced Human HSPCs

1. Replication-defective lentiviral vector particles (95): Produce and titer as described in other Chapters of this book.

2. Human CD34$^+$ cord blood cells (StemCell Technologies, Vancouver, BC, Canada).

3. Human HSPC culture medium: IMDM containing 10% FBS, 100 μM β-mercaptoethanol (Sigma-Aldrich Corp.), plus human stem cell factor (SCF; 100 ng/mL), human Flt-3 ligand (100 ng/mL), and human thrombopoietin (20 ng/mL).

4. Recombinant fibronectin fragment (RetroNectin; Takara Mirus Bio, Madison, WI, USA).

5. Protamine sulfate solution (1000×; 4 mg/ml) (Sigma-Aldrich Corp.). Prepare stock of 4 mg/ml (1000×) in sterile deionized, distilled water; aliquot and store at −20°C.

6. Human HSPC differentiation medium: IMDM medium containing 10% heat-inactivated FBS, L-glutamine (2 mM), and penicillin (50 IU/mL), plus human interleukin-3 (IL-3; 20 ng/mL), human interleukin-6 (IL-6; 20 ng/mL) and human granulocyte-macrophage colony-stimulating factor (20 ng/mL).

7. Antibodies: anti-CD34 and anti-CD14 monoclonal antibodies conjugated to allophycocyanin (BD Biosciences).

8. Analytical flow cytometer.

2.4.4. Transgene Expression Following ESC Differentiation

1. Replication-defective lentiviral vector particles (95): Produce and titer as described in other chapters of this book.

2. ESC line adapted for growth on gelatin-coated culture dishes (e.g., CCE (16); StemCell Technologies, cat. no. 00300).

3. ESC culture medium: Grow undifferentiated CCE ESCs on 0.1% gelatin-coated dishes in DMEM with 4.5 g/l glucose and 4 mM L-glutamine plus 150 μM β-mercaptoethanol, 15% FBS, 10 ng/mL murine leukemia inhibitory factor (LIF; StemCell Technologies).

4. ESC embryoid body differentiation medium: IMDM containing 1% methylcellulose, 2 mM l-glutamine, $150\,\mu M$ β-mercaptoethanol, 15% FBS, and 40 ng/ml murine SCF (StemCell Technologies).

5. ESC hematopoietic differentiation medium: IMDM with 1% methylcellulose, 2 mM l-glutamine, 1% bovine serum albumin, $10\,\mu g/ml$ insulin, $200\,\mu g/ml$ transferrin, $150\,\mu M$ β-mercaptoethanol, 15% FBS, 150 ng/ml SCF, 30 ng/ml murine IL-3, 30 ng/ml murine IL-6, and 3 U/ml erythropoietin (StemCell Technologies).

6. Monoclonal antibodies: anti-CD41 and anti-CD45 conjugated to R-phycoerythrin (BD Biosciences).

7. Analytical flow cytometer.

2.4.5. Murine HSPC Immortalization Assay

1. Female C57BL/6 mice (6- to 8-week old; The Jackson Laboratory, Bar Harbor, ME, USA; cat. no. 000664) used as bone marrow donors. All procedures involving mice must follow the guidelines set forth in the National Institutes of Health Guide for the Care and Use of Laboratory Animals and be approved by an Institutional Animal Care and Use Committee.

2. 5-fluorouracil (5-FU; Sigma-Aldrich Corp., cat. no. F6627). Store the stock at room temperature, avoiding light. Prepare a fresh working solution of 15 mg/ml in PBS immediately before use.

3. Erythrocyte lysing solution: 154 mM NH_4Cl 10 mM $NaHCO_3$, and 0.082 mM sodium ethylenediaminetetraacetic acid (EDTA), pH 7.3. Commercial lysing solutions are also available. Store at room temperature.

4. Murine HSPC culture medium: IMDM supplemented with 4.5 g/l glucose, 2 mM L-glutamine, 50 IU/ml penicillin, 50 µg/ml streptomycin, 15% heat-inactivated FBS, 100 ng/ml murine SCF, 30 ng/ml murine IL-3, and 10 ng/ml murine IL-6. Store at 4°C and warm up to 37°C before use.

5. Recombinant fibronectin fragment (RetroNectin; Takara Mirus Bio).

6. Polybrene stock solution (1000×; 8 mg/ml) prepared as described in Subheading 2.1.3. Aliquot and store at –20°C.

3. Methods

3.1. Barrier Function Assay

As discussed in Subheading 1.3, one of the defining properties of an insulator is to function as a barrier to prevent adjacent heterochromatin from advancing into and silencing gene expression. The barrier function assay will determine if the test element can

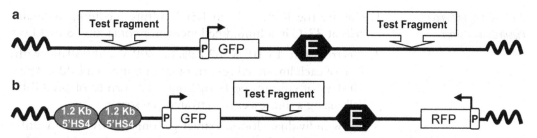

Fig. 2. Schematic diagram showing the reporter plasmid pRL-TKGFP-CMVE used in the flow cytometry-based barrier function assay (**a**) and the reporter plasmid pRL-TKGFP-CMVE-TKRFP used in the flow cytometry-based enhancer blocking assay (**b**). *1.2 kb 5'HS4* 1.2-kb fragment of the chicken β-globin 5'HS4 insulator, *GFP* enhanced GFP gene, *RFP* DsRed.T4 gene, *E* human CMV immediate early region enhancer, *P* herpes simplex virus TK promoter

reliably shield a stably integrated transgene from chromosomal position effects and allow it to be expressed in a copy number-dependent manner (Fig. 2a). In the absence of an insulator, transgene expression tends to be copy number-independent and is frequently extinguished in cells over time; when shielded with an insulator, transgene expression should be stable with a good correlation between vector copy number and expression levels. Felsenfeld and colleagues established an assay to test the barrier activity of the 1.2-kb fragment of the chicken β-globin 5'HS4 insulator (68). A reporter expressing a portion (Tac subunit) of the human IL-2 receptor driven by an erythroid-specific promoter and enhancer was constructed and integrated into a chicken erythroid cell line. In the absence of the insulator, extended culture of the transfected cells often led to loss of transgene expression, reflecting integration site position effects. When the same reporter plasmid was flanked on each side by two direct copies of the 1.2-kb 5'HS4 insulator and used to transfect the cells, stable transgene expression was obtained even after extended period of time in culture. The following is a modified version of the above barrier function assay that takes advantage of the intrinsic fluorescence of the GFP reporter gene in order to facilitate generation of single-copy clones and analyze transgene expression levels by fluorescence-activated cell sorting and flow cytometry.

3.1.1. Design and Construction of Reporter Plasmids for Barrier Function Assay

1. The reporter plasmid pRL-TKGFP-CMVE (Fig. 2a) can be used to determine the barrier function activity of any test fragments with insulator properties.

2. Subclone one or more copies of the test element with potential insulating activity upstream of the TK-GFP expression cassette and an equivalent number of copies of the test element downstream of the CMV enhancer (see Note 3).

3. Construct a positive control insulated reporter plasmid by removing two direct copies of the full-length 1.2-kb chicken β-globin 5'HS4 insulator from plasmid pJC13-1 and inserting them upstream of the TK-GFP expression cassette and downstream of the CMV enhancer.

1. Culture the K562 cells in K562 culture medium. Maintain cells at 37°C in a humidified incubator containing 5% CO_2.

2. Electroporate 1×10^7 K562 cells (at 300 V and 950 μF) with 1 μg of each linearized reporter plasmids in 0.5 ml PBS. After 10 days of culture, perform single cell sorting of the GFP$^+$ cells using a fluorescence-activated cell sorter.

3. Grow individual clones, extract genomic DNA, and identify clones with single-copy integrations by Southern blot analysis. Digest the genomic DNA with enzymes that cut only once within the reporter plasmid, hybridize Southern blots with a GFP-specific probe, and pick clones that exhibit a single band.

4. Perform flow cytometric analysis on single-copy clones immediately and at one month intervals for up to 3 months of continuous culture. Without any insulators, many clones will express little or no GFP. Also, many of the clones that are initially GFP$^+$ will gradually extinguish expression over time. Expect most or all of the single-copy clones to express and maintain GFP expression in the presence of full insulator activity (see Note 4).

3.2. Enhancer Blocking
Assay

The enhancer blocking assay is used to measure the ability of a test element to shield a promoter from an enhancer when it is placed between the enhancer and a promoter (66, 96) (Fig. 1b). Chung *et al.* developed an enhancer blocking assay to test the ability of the 1.2-kb chicken β-globin 5′HS4 insulator and shield a reporter expressing a neomycin (*neo*) resistance gene from the action of a strong enhancer (66). An erythroid-specific enhancer and promoter were used to drive *neo* gene expression in stably transfected human K562 erythroleukemia cells and geneticin-resistant colonies were counted. Using this assay, it was shown that the 1.2-kb 5′HS4 fragment was effective in reducing the number of drug-resistant colonies, which was an indication of the strength of enhancer blocking activity.

We have recently developed a flow cytometry-based assay (92) by constructing a plasmid in which enhancer blocking activity is measured by down-modulation of expression of a GFP reporter gene when a test fragment is inserted between a herpes simplex virus TK promoter-driven GFP expression cassette and a downstream human CMV enhancer (Fig. 2b). To control for variations in transfection efficiency, GFP expression levels are compared with those of an analogous CMV enhancer-TK promoter-driven RFP reporter gene expression cassette contained in the same construct. To allow use of circular plasmids in transient transfection experiments and eliminate influences of flanking chromatin structure on GFP expression in stable transfection experiments with linearized plasmids, two copies of the full-length 1.2-kb 5′HS4 insulator were introduced upstream of the GFP expression cassette. The various control and test constructs are transiently transfected into human 293T cells or stably transfected into human K562 erythroleukemia cells. Forty-eight hours following transient transfection of 293T cells and 2–4 weeks after stable transfection of K562 cells, GFP fluorescence can be measured by flow cytometric analysis and normalized with respect to RFP fluorescence.

3.2.1. Design and Construction of Reporter Plasmids for Enhancer Blocking Assay

1. The reporter plasmid pRL-TKGFP-CMVE-TKRFP (Fig. 2b) can be used to determine the enhancer blocking activity of any test fragments with insulator properties.

2. Subclone one or more copies of the test element with potential enhancer blocking activity between the TK-GFP expression cassette and the CMV enhancer at a unique *Cla*I restriction enzyme site.

3. Construct a reporter plasmid to control for spacing between the TK promoter of the TK-GFP expression cassette and the CMV enhancer by inserting a fragment of bacteriophage λ DNA equivalent in size to the test fragment at the unique *Cla*I restriction enzyme site.

4. For comparison, construct a positive control insulated reporter plasmid by inserting one or two copies of the full-length 1.2-kb chicken β-globin 5′HS4 insulator between the TK-GFP expression cassette and the CMV enhancer at the unique *Cla*I restriction enzyme site.

3.2.2. Testing of Reporter Plasmids for Enhancer Blocking Activity

1. Culture the 293T and the K562 cells in their corresponding culture media. Maintain cells at 37°C in a humidified incubator containing 5% CO_2.

2. Transiently transfect 293T cells with plasmid DNA (20 μg) by calcium phosphate coprecipitation. Forty-eight hours following the transient transfection of 293T cells, measure GFP fluorescence by flow cytometric analysis and normalize with respect to RFP expression levels.

3. Electroporate 1×10^7 K562 cells (at 300V and 950μF) using 20 μg of each linearized reporter plasmids in 0.5 ml PBS. After 10 days of culture, perform single cell sorting of the GFP+ cells using a fluorescence-activated cell sorter. Two to four weeks after stable transfection of K562 cells, measure GFP fluorescence by flow cytometric analysis and normalize with respect to RFP fluorescence intensity. Obtain an indication of the strength of the enhancer blocking activity of the test fragment based on the relative decrease in GFP expression.

3.3. Lentiviral Vector Construction

3.3.1. Strong Enhancer–Promoter Sequences

SIN HIV-1-based lentiviral transfer vectors utilizing constitutively active viral and cellular promoters to drive expression of the GFP reporter gene have been described previously (43, 50). Examples of promoters are given below that could be utilized as internal promoters in any SIN lentiviral vector (see Note 5).

1. MSCV LTR promoter: A 302-bp *Nhe*I–*Sma*I fragment of the U3 region and part of the R region of the MSCV LTR (nucleotides –266 to +30 relative to the U3/R boundary) can be obtained from the MSCV gammaretroviral vector (17).

2. Human EF1α gene promoter: A 1.2-kb *Hin*dIII-*Eco*RI fragment, which includes 203 bp of 5′ flanking region, 33 bp of the first exon, 943 bp of the first intron, and 10 bp of the second exon (located 20 bp upstream of the EF1α ATG initiation codon), can be obtained from the plasmid pEF-BOS (39).

3. CAG promoter: A 1.8-kb *Sph*I-*Sma*I fragment, which contains 365 bp of the CMV enhancer, 277 bp of the chicken β-actin promoter and a 908-bp hybrid intron with a 3′ splice site from the rabbit β-globin gene, can be excised from the pBacMam-2 plasmid.

4. Subclone different promoter fragments into the lentiviral vector upstream of the transgene or a selectable marker.

5. Prepare a high-quality plasmid DNA for transient transfection and generation of vector particles.

3.3.2. S/MAR Element

One of the best characterized S/MAR elements is a 2.2-kb fragment (element E) from the upstream border of the human interferon-β (IFN-β) gene, of which an 800-bp core (IFN-SAR) contributes most of the S/MAR activity (52, 55, 56).

1. Obtain the IFN-SAR as an 800-bp fragment from plasmid pCL and clone in reverse orientation into the 3′ untranslated region of the transgene, immediately upstream of the 3′ LTR.

2. Prepare a high-quality plasmid DNA for transient transfection and generation of vector particles.

3.3.3. 5′ HS4 Insulator

1. The 5′HS4-containing vectors can be constructed by inserting the 1.2-kb fragment of pJC13-1 plasmid containing the chicken β-globin 5′HS4 insulator in direct orientation into the deleted U3 region of the 3′ LTR.

2. Prepare a high-quality plasmid DNA for transient transfection and generation of vector particles.

3.4. Lentiviral Vector Function

A convenient strategy to flank the lentiviral transgene expression cassette with a boundary element is to subclone the fragment of interest into the U3 region of the 3′ LTR of the vector, which is

transferred to the 5′ LTR during viral replication. A potential
limitation of this approach (e.g., in the case of an element the size
of the 1.2-kb chicken β-globin 5′HS4 insulator) is a decrease in
vector titer, perhaps due to topological or size constraints (50).
As full insulator activity may require multiple copies of the element in question (66), another potential cause for concern in
addition to size constraints is that direct repeats are frequently
unstable during viral replication (32). Besides a flanking configuration (55, 57), a single copy of an S/MAR element (the IFN-
SAR) has been shown to provide beneficial effects (50, 55,
60–62), so this might be an option in certain instances. Other
elements have been described that have been reported to help
counteract insertion site-associated chromosomal position effects
when used as a single copy in the context of a lentiviral vector,
such as locus control regions (LCRs) and ubiquitously acting
chromatin opening elements (UCOEs) (97, 98). However, their
generally large size (>2 kb) could also pose size constraints as far
as titer is concerned (83). Therefore, when incorporating any
boundary or novel genetic regulatory elements into lentiviral vectors, especially multiple copies, it is absolutely essential to confirm
the fidelity of vector sequence transmission while concurrently
assessing vector performance.

*3.4.1. Transgene
Expression in Cell Lines*

1. To evaluate the activity of S/MAR and insulator elements in
   the context of lentiviral vectors, produce replication-defective
   vector particles and measure titers. Compare the titers
   between the parental vector and the vectors containing the
   various elements by flow cytometric analysis of GFP
   fluorescence.

2. To assess vector performance, transduce K562 cells using low
   multiplicities of infection (MOI) and compare the levels of
   GFP transgene expression achieved by the vectors containing
   the various elements to those achieved with the parental vector by flow cytometric analysis of GFP fluorescence. Pooled
   populations or clones containing single-copy integrations
   (obtained by fluorescence-activated cell sorting and Southern
   blot analysis) can be followed initially and at 1 month intervals for up to 3 months of continuous culture.

*3.4.2. Stability of the
Inserted Elements*

1. To document proper transmission to the 5′ LTR and stability
   of insulators incorporated into the 3′ LTR, extract genomic
   DNA from vector-transduced cells.

2. Use primer pairs designed to specifically PCR-amplify the 5′
   and 3′ LTR sequences separately and analyze for the presence
   of DNA bands having the expected sizes (see Note 6).

3. Sequence the DNA fragments to further confirm the structurally intact presence of the insulator following transmission
   from the 3′ LTR to the 5′ LTR.

3.4.3. Transgene Expression Following Differentiation of Transduced Human HSPCs

The effects of S/MAR and insulator elements on transgene expression can also be studied in primary human myelomonocytic cells following transduction and in vitro differentiation of human CD34$^+$ HSPCs.

1. Obtain human CD34$^+$ cord blood cells and culture in 12-well tissue culture plates coated with recombinant fibronectin fragment ($2\,\mu g/cm^2$) at a density of 1×10^6 cells per well. Prestimulate the cells for 48 h in human HSPC culture medium.

2. Transduce the cells for 24 h with lentiviral vector particles (2×10^6 transducing units (TU)/mL; MOI, 2) in the presence of protamine sulfate ($4\,\mu g/mL$). Add fresh medium and culture the cells for an additional 72 h.

3. To evaluate and compare the levels of GFP transgene expression in CD14$^+$ monocytes derived from transduced human CD34$^+$ cord blood cells, culture the cells for 7 weeks in human HSPC differentiation medium.

4. Maintain all cultures at 37°C in a humidified atmosphere containing 5% CO_2. At weekly intervals, stain cells with anti-CD14 monoclonal antibodies conjugated to allophycocyanin and measure GFP fluorescence in CD14$^+$ cells by flow cytometric analysis (see Note 7).

3.4.4. Transgene Expression Following ESC Differentiation

Most gammaretroviral vectors experience severe downregulation of transgene expression upon differentiation of ESCs into more mature cells in vitro (32). As noted in the Subheading 1, when assessed at the single vector copy per cell level, transgenes delivered by lentiviral vectors have also been found to be susceptible to variegated expression and silencing in ESCs and their differentiated progeny (23, 26, 29). Consequently, it is important to verify efficacy of lentiviral vector-directed transgene expression in differentiating ESC systems prior to gene function studies. The example provided involves differentiation of ESCs into hematopoietic cells.

1. Prepare lentiviral vector particles and transduce monolayers of CCE ESCs cells in the presence of polybrene at $8\,\mu g/ml$.

2. Determine stability of transgene expression by measuring GFP fluorescence by flow cytometric analysis after 10 days and 1 month of continuous culture.

3. Use a two-step method for in vitro hematopoietic differentiation of CCE ESCs (16). The primary differentiation of ESCs into embryoid bodies can be induced by removal of LIF and plating a single-cell suspension of cells in ESC embryoid body differentiation medium.

4. Perform hematopoietic differentiation of embryoid bodies by plating single-cell suspensions of day 9 embryoid bodies in ESC hematopoietic differentiation medium.

5. Stain differentiated cells with anti-CD41 and anti-CD45 monoclonal antibodies and measure GFP fluorescence of CD41$^+$ and/or CD45$^+$ cells by flow cytometric analysis.

3.4.5. Murine HSPC Immortalization Assay

1. Dissolve 5-FU in sterile PBS (15 mg/ml) immediately before use and intravenously inject into mice (150 mg/kg body weight) using a 27-G needle attached to a 1-ml syringe.

2. Harvest HSPC-enriched bone marrow cells 4 days after the 5-FU injection. Flush the hind limbs with PBS containing 2% FBS using a 21-G needle attached to a 5-ml syringe (see Note 8).

3. Lyse erythrocytes by incubating total bone marrow cells in erythrocyte lysis solution for 10 min at room temperature followed by centrifugation at 375×g for 5 min.

4. Coat 35-mm suspension culture plates with 2 μg/cm^2 recombinant fibronectin fragment. Transfer the nucleated cells to plates at a density of 5×10^5 cells/ml and culture for 48 h in murine HSPC culture medium at 37°C in a humidified atmosphere containing 5% CO_2.

5. Transduce the cells for 3 consecutive days (4 h each day) by incubation with vector conditioned medium in the presence of 8 μg/ml polybrene supplemented with 100 ng/ml murine SCF, 30 ng/ml murine IL-3, and 10 ng/ml murine IL-6.

6. Culture the transduced cells in murine HSPC culture medium minus murine SCF and murine IL-6. Passage cells every 3 days for 2 weeks. After bulk culture, plate 100 cells into each well of a 96-well plate and continue culturing them in HSPC culture medium minus murine SCF and murine IL-6. Two weeks later, examine the plates and determine the replating frequency based on the number of wells that contain proliferating cell populations.

7. Expand proliferating cell populations to establish immortalized cell lines.

8. Extract genomic DNA from the immortalized cells and subject to Southern blot analysis to determine vector copy number. Perform integration site analysis (e.g., by linear amplification-mediated polymerase chain reaction (LAM-PCR)).

4. Notes

1. The 293T/17 cell line is a clone of the 293T (293tsA-1609neo) human embryonic kidney cell line (99), which was selected for high transfectability and capability of producing high-titer vector stocks. The 293T cell line is a derivative of 293 cells into which the simian virus 40 (SV40) large T antigen gene was inserted. 293 cells express the adenovirus serotype 5 E1A 12S and 13S gene products, which strongly transactivate transcription from expression vectors containing the human CMV enhancer-promoter (100). Expression of the SV40 large T antigen by 293T cells may stimulate extra-chromosomal replication of plasmids containing the SV40 origin of replication during transient transfection.

2. Besides academic sources, lentiviral vector backbones and packaging systems are available commercially (e.g., Clontech Laboratories, Inc., Mountain View, CA, USA; Invitrogen Corp., Carlsbad, CA, USA; Stratagene Corp., La Jolla, CA, USA).

3. Common recombinant DNA techniques are used for plasmid DNA preparations, restriction enzyme digestions and subclonings. Detailed protocols for each technique can be obtained from commercial sources, as well as from various standard molecular biology manuals. Accordingly, these techniques have not been described here.

4. When GFP is used as the reporter gene (101), it is useful to wait ~5 days before analyzing the transduced cells by flow cytometry. This will minimize the contribution of false-positive signals because of pseudotransduction, which is the direct transfer of reporter protein present in the vector supernatants or incorporated into the vector particles to the target cells; this is particularly problematic for vesicular stomatitis virus-G-protein pseudotyped vectors (102, 103). Note that it has also been shown that transgenes can be efficiently transiently expressed from unintegrated lentiviral vectors during this timeframe (104).

5. Both the EF1α and the CAG promoters contain introns for augmented transgene expression (38, 40). Consequently, they are relatively large in size (1.2 and 1.8 kb, respectively). Size constraints and the fact that introns in gammatroviruses are removed during the viral life cycle (105) have precluded routine use of these promoters in gammaretroviral vectors. Lentiviral vectors based on HIV-1 are capable of transporting intron-containing sequences to target cells through a process

that involves the binding of the viral Rev protein to a *cis*-acting RNA sequence known as the Rev-responsive element (106).

6. The presence of shorter or longer PCR products than expected indicates deletions or rearrangements of the inserted elements.

7. At the 7-week time point in the experiment, ~30–50% of the nonadherent cells should be CD14⁺ monocytes. Detailed protocols for staining of hematopoietic cells and detection of cell surface antigens by immunofluorescence flow cytometric analysis may be found on the websites of the monoclonal antibody manufacturers.

8. It should be possible to obtain $3–4 \times 10^6$ bone marrow cells from each 5-FU-treated mouse. The bone marrow cells can be used directly or further enriched for more primitive HSPCs using various magnetic- or fluorescence-activated cell sorting procedures (107). When culturing murine HSPCs, keep the cell density at 0.5×10^6 cells/ml.

Acknowledgments

This work was supported in part by National Institutes of Health grants R01HL65519 and R01HL66305, and by an Elaine H. Snyder Cancer Research Award and a King Fahd Endowed Professorship (to R.G.H.) from The George Washington University Medical Center.

References

1. Hawley, R. G. (2001) Progress toward vector design for hematopoietic stem cell gene therapy. *Curr. Gene Ther.* **1**, 1–17.

2. Miller, D. G., Adam, M. A., and Miller, A. D. (1990) Gene transfer by retrovirus vectors occurs only in cells that are actively replicating at the time of infection. *Mol. Cell. Biol.* **10**, 4239–42.

3. Lewis, P. F. and Emerman, M. (1994) Passage through mitosis is required for oncoretroviruses but not for the human immunodeficiency virus. *J. Virol.* **68**, 510–6.

4. Yamashita, M., Perez, O., Hope, T. J., and Emerman, M. (2007) Evidence for direct involvement of the capsid protein in HIV infection of nondividing cells. *PLoS Pathog.* **3**, 1502–10.

5. Naldini, L., Blomer, U., Gallay, P., Ory, D., Mulligan, R., Gage, F. H. *et al.* (1996) *In vivo* gene delivery and stable transduction of nondividing cells by a lentiviral vector. *Science* **272**, 263–7.

6. Ramezani, A. and Hawley, R. G. (2002) Overview of the HIV-1 lentiviral vector system. *Curr. Protoc. Mol. Biol.* **16.21**, 1–15.

7. Speers, W. C., Gautsch, J. W., and Dixon, F. J. (1980) Silent infection of murine embryonal carcinoma cells by Moloney murine leukemia virus. *Virology* **105**, 241–4.

8. Jahner, D., Stuhlmann, H., Stewart, C. L., Harbers, K., Lohler, J., Simon, I. *et al.* (1982) De novo methylation and expression of retroviral genomes during mouse embryogenesis. *Nature* **298**, 623-8.

9. Niwa, O., Yokota, Y., Ishida, H., and Sugahara, T. (1983) Independent mechanisms involved in suppression of the Moloney leukemia virus genome during differentiation of murine teratocarcinoma cells. *Cell* **32**, 1105–13.

10. Hilberg, F., Stocking, C., Ostertag, W., and Grez, M. (1987) Functional analysis of a retroviral host-range mutant: altered long terminal repeat sequences allow expression in embryonal carcinoma cells. *Proc. Natl. Acad. Sci. USA* **84**, 5232–6.

11. Weiher, H., Barklis, E., Ostertag, W., and Jaenisch, R. (1987) Two distinct sequence elements mediate retroviral gene expression in embryonal carcinoma cells. *J. Virol.* **61**, 2742–6.

12. Barklis, E., Mulligan, R. C., and Jaenisch, R. (1986) Chromosomal position or virus mutation permits retrovirus expression in embryonal carcinoma cells. *Cell* **47**, 391–9.

13. Petersen, R., Kempler, G., and Barklis, E. (1991) A stem cell specific silencer in the primer-binding site of a retrovirus. *Mol. Cell. Biol.* **11**, 1214–21.

14. Wolf, D. and Goff, S. P. (2007) TRIM28 mediates primer binding site-targeted silencing of murine leukemia virus in embryonic cells. *Cell* **131**, 46–57.

15. Grez, M., Akgün, E., Hilberg, F., and Ostertag, W. (1990) Embryonic stem cell virus, a recombinant murine retrovirus with expression in embryonic stem cells. *Proc. Natl. Acad. Sci. USA* **87**, 9202–6.

16. Keller, G., Wall, C., Fong, A. Z. C., Hawley, T. S., and Hawley, R. G. (1998) Overexpression of HOX11 leads to the immortalization of embryonic precursors with both primitive and definitive hematopoietic potential. *Blood* **92**, 877–87.

17. Hawley, R. G., Lieu, F. H. L., Fong, A. Z. C., and Hawley, T. S. (1994) Versatile retroviral vectors for potential use in gene therapy. *Gene Ther.* **1**, 136–8.

18. Hawley, R. G., Hawley, T. S., Fong, A. Z. C., Quinto, C., Collins, M., Leonard, J. P. et al. (1996) Thrombopoietic potential and serial repopulating ability of murine hematopoietic stem cells constitutively expressing interleukin-11. *Proc. Natl. Acad. Sci. USA* **93**, 10297–302.

19. Henikoff, S. (1992) Position effect and related phenomena. *Curr. Opin. Genet. Dev.* **2**, 907–12.

20. Rivella, S. and Sadelain, M. (1998) Genetic treatment of severe hemoglobinopathies: the combat against transgene variegation and transgene silencing. *Semin. Hematol.* **35**, 112–25.

21. Emery, D. W. and Stamatoyannopoulos, G. (1999) Stem cell gene therapy for the β-chain hemoglobinopathies. Problems and progress. *Ann. N. Y. Acad. Sci.* **872**, 94–107.

22. Talbert, P. B. and Henikoff, S. (2006) Spreading of silent chromatin: inaction at a distance. *Nat. Rev. Genet.* **7**, 793–803.

23. Pannell, D., Osborne, C. S., Yao, S., Sukonnik, T., Pasceri, P., Karaiskakis, A. et al. (2000) Retrovirus vector silencing is de novo methylase independent and marked by a repressive histone code. *EMBO J.* **19**, 5884–94.

24. Jordan, A., Defechereux, P., and Verdin, E. (2001) The site of HIV-1 integration in the human genome determines basal transcriptional activity and response to Tat transactivation. *EMBO J.* **20**, 1726–38.

25. Pawliuk, R., Westerman, K. A., Fabry, M. E., Payen, E., Tighe, R., Bouhassira, E. E. et al. (2001) Correction of sickle cell disease in transgenic mouse models by gene therapy. *Science* **294**, 2368–71.

26. Lois, C., Hong, E. J., Pease, S., Brown, E. J., and Baltimore, D. (2002) Germline transmission and tissue-specific expression of transgenes delivered by lentiviral vectors. *Science* **295**, 868–72.

27. Persons, D. A., Hargrove, P. W., Allay, E. R., Hanawa, H., and Nienhuis, A. W. (2003) The degree of phenotypic correction of murine β-thalassemia intermedia following lentiviral-mediated transfer of a human γ-globin gene is influenced by chromosomal position effects and vector copy number. *Blood* **101**, 2175–83.

28. Yao, S., Sukonnik, T., Kean, T., Bharadwaj, R. R., Pasceri, P., and Ellis, J. (2004) Retrovirus silencing, variegation, extinction, and memory are controlled by a dynamic interplay of multiple epigenetic modifications. *Mol. Ther.* **10**, 27–36.

29. Ellis, J. (2005) Silencing and variegation of gammaretrovirus and lentivirus vectors. *Hum. Gene Ther.* **16**, 1241–6.

30. Tamkun, J. W., Deuring, R., Scott, M. P., Kissinger, M., Pattatucci, A. M., Kaufman, T. C. et al. (1992) Brahma: a regulator of Drosophila homeotic genes structurally related to the yeast transcriptional activator SNF2/SWI2. *Cell* **68**, 561–72.

31. Aparicio, O. M. and Gottschling, D. E. (1994) Overcoming telomeric silencing: a trans-activator competes to establish gene expression in a cell cycle-dependent way. *Genes Dev.* **8**, 1133–46.

32. Ramezani, A., Hawley, T. S., and Hawley, R. G. (2006) Stable gammaretroviral vector expression during embryonic stem cell-derived *in vitro* hematopoietic development. *Mol. Ther.* **14**, 245–54.

33. Forrester, W. C., Thompson, C., Elder, J. T., and Groudine, M. (1986) A developmentally stable chromatin structure in the human β-globin gene cluster. *Proc. Natl. Acad. Sci. USA* **83**, 1359–63.

34. Jenuwein, T., Forrester, W. C., Qiu, R. G., and Grosschedl, R. (1993) The immunoglobulin mu enhancer core establishes local factor

access in nuclear chromatin independent of transcriptional stimulation. *Genes Dev.* **7**, 2016–32.

35. Pikaart, M., Feng, J., and Villeponteau, B. (1992) The polyomavirus enhancer activates chromatin accessibility on integration into the *HPRT* gene. *Mol. Cell Biol.* **12**, 5785–92.

36. Walters, M. C., Magis, W., Fiering, S., Eidemiller, J., Scalzo, D., Groudine, M. *et al.* (1996) Transcriptional enhancers act in cis to suppress position-effect variegation. *Genes Dev.* **10**, 185–95.

37. Francastel, C., Walters, M. C., Groudine, M., and Martin, D. I. (1999) A functional enhancer suppresses silencing of a transgene and prevents its localization close to centromeric heterochromatin. *Cell* **99**, 259–69.

38. Kim, D. W., Uetsuki, T., Kaziro, Y., Yamaguchi, N., and Sugano, S. (1990) Use of the human elongation factor 1α promoter as a versatile and efficient expression system. *Gene* **91**, 217–23.

39. Mizushima, S. and Nagata, S. (1990) pEF-BOS, a powerful mammalian expression vector. *Nucl. Acids Res.* **18**, 5322.

40. Niwa, H., Yamamura, K., and Miyazaki, J. (1991) Efficient selection for high-expression transfectants with a novel eukaryotic vector. *Gene* **108**, 193–9.

41. Cheng, L., Du, C., Lavau, C., Chen, S., Tong, J., Chen, B. P. *et al.* (1998) Sustained gene expression in retrovirally transduced, engrafting human hematopoietic stem cells and their lympho-myeloid progeny. *Blood* **92**, 83–92.

42. Dorrell, C., Gan, O. I., Pereira, D. S., Hawley, R. G., and Dick, J. E. (2000) Expansion of human cord blood CD34+CD38- cells in *ex vivo* culture during retroviral transduction without a corresponding increase in SCID repopulating cell (SRC) frequency: dissociation of SRC phenotype and function. *Blood* **95**, 102–10.

43. Ramezani, A., Hawley, T. S., and Hawley, R. G. (2000) Lentiviral vectors for enhanced gene expression in human hematopoietic cells. *Mol. Ther.* **2**, 458–69.

44. Gao, Z., Golob, J., Tanavde, V. M., Civin, C. I., Hawley, R. G., and Cheng, L. (2001) High levels of transgene expression following transduction of long-term NOD/SCID-repopulating human cells with a modified lentiviral vector. *Stem Cells* **19**, 247–59.

45. Taboit-Dameron, F., Malassagne, B., Viglietta, C., Puissant, C., Leroux-Coyau, M., Chereau, C. *et al.* (1999) Association of the 5'HS4 sequence of the chicken β-globin locus control region with human EF1α gene promoter induces ubiquitous and high expression of human CD55 and CD59 cDNAs in transgenic rabbits. *Transgenic Res.* **8**, 223–35.

46. Chang, L.-J., Urlacher, V., Iwakuma, T., Cui, Y., and Zucali, J. (1999) Efficacy and safety analyses of a recombinant human immunodeficiency virus type 1 derived vector system. *Gene Ther.* **6**, 715–28.

47. Ye, Z.-Q., Qui, P., Burkholder, J. K., Turner, J., Culp, J., Roberts, T. *et al.* (1998) Cytokine transgene expression and promoter usage in primary CD34+ cells using particle-mediated gene delivery. *Hum. Gene Ther.* **9**, 2197–205.

48. Araki, K., Imaizumi, T., Okuyama, K., Oike, Y., and Yamamura, K. (1997) Efficiency of recombination by Cre transient expression in embryonic stem cells: comparison of various promoters. *J. Biochem.* **122**, 977–82.

49. Okabe, M., Ikawa, M., Kominami, K., Nakanishi, T., and Nishimune, Y. (1997) 'Green mice' as a source of ubiquitous green cells. *FEBS Let.* **407**, 313–9.

50. Ramezani, A., Hawley, T. S., and Hawley, R. G. (2003) Performance- and safety-enhanced lentiviral vectors containing the human interferon-β scaffold attachment region and the chicken β-globin insulator. *Blood* **101**, 4717–24.

51. Bode, J. and Maass, K. (1988) Chromatin domain surrounding the human interferon-β gene as defined by scaffold-attached regions. *Biochemistry* **27**, 4706–11.

52. Mielke, C., Kohwi, Y., Kohwi-Shigematsu, T., and Bode, J. (1990) Hierarchical binding of DNA fragments derived from scaffold-attached regions: correlation of properties *in vitro* and function *in vivo*. *Biochemistry* **29**, 7475–85.

53. Bode, J., Kohwi, Y., Dickinson, L., Joh, T., Klehr, D., Mielke, C. *et al.* (1992) Biological significance of unwinding capability of nuclear matrix-associating DNAs. *Science* **255**, 195–7.

54. Boulikas, T. (1993) Nature of DNA sequences at the attachment regions of genes to nuclear matrix. *J. Cell. Biochem.* **52**, 14–22.

55. Schubeler, D., Mielke, C., Maass, K., and Bode, J. (1996) Scaffold/mattrix-attached regions act upon transcription in a context-dependent manner. *Biochemistry* **35**, 11160–9.

56. Benham, C., Kohwi-Shigematsu, T., and Bode, J. (1997) Stress-induced duplex DNA destabilization in scaffold/matrix attachment regions. *J. Mol. Biol.* **274**, 181–96.

57. Goetze, S., Baer, A., Winkelmann, S., Nehlsen, K., Seibler, J., Maass, K. *et al.* (2005) Performance of genomic bordering elements at predefined genomic loci. *Mol. Cell. Biol.* **25**, 2260–72.

58. Forrester, W. C., Fernandez, L. A., and Grosschedl, R. (1999) Nuclear matrix attachment regions antagonize methylation-dependent repression of long-range enhancer-promoter interactions. *Genes Dev.* **13**, 3003–14.

59. Fernandez, L. A., Winkler, M., and Grosschedl, R. (2001) Matrix attachment region-dependent function of the immunoglobulin μ enhancer involves histone acetylation at a distance without changes in enhancer occupancy. *Mol. Cell. Biol.* **21**, 196–208.

60. Agarwal, M., Austin, T. W., Morel, F., Chen, J., Bohnlein, E., and Plavec, I. (1998) Scaffold attachment region-mediated enhancement of retroviral vector expression in primary T cells. *J. Virol.* **72**, 3720–8.

61. Auten, J., Agarwal, M., Chen, J., Sutton, R., and Plavec, I. (1999) Effect of scaffold attachment region on transgene expression in retrovirus vector-transduced primary T cells and macrophages. *Hum. Gene Ther.* **10**, 1389–99.

62. Dang, Q., Auten, J., and Plavec, I. (2000) Human β interferon scaffold attachment region inhibits de novo methylation and confers long-term, copy number-dependent expression to a retroviral vector. *J. Virol.* **74**, 2671–8.

63. Park, F. and Kay, M. A. (2001) Modified HIV-1 based lentiviral vectors have an effect on viral transduction efficiency and gene expression *in vitro* and *in vivo*. *Mol. Ther.* **4**, 164–73.

64. Bushey, A. M., Dorman, E. R., and Corces, V. G. (2008) Chromatin insulators: regulatory mechanisms and epigenetic inheritance. *Mol. Cell* **32**, 1–9.

65. Wallace, J. A. and Felsenfeld, G. (2007) We gather together: insulators and genome organization. *Curr. Opin. Genet. Dev.* **17**, 400–7.

66. Chung, J. H., Whiteley, M., and Felsenfeld, G. (1993) A 5′ element of the chicken β-globin domain serves as an insulator in human erythroid cells and protects against position effect in *Drosophila*. *Cell* **74**, 505–14.

67. Chung, J. H., Bell, A. C., and Felsenfeld, G. (1997) Characterization of the chicken β-globin insulator. *Proc. Natl. Acad. Sci. USA* **94**, 575–80.

68. Pikaart, M. J., Recillas-Targa, F., and Felsenfeld, G. (1998) Loss of transcriptional activity of a transgene is accompanied by DNA methylation and histone deacetylation and is prevented by insulators. *Genes Dev.* **12**, 2852–62.

69. Bell, A. C., West, A. G., and Felsenfeld, G. (1999) The protein CTCF is required for the enhancer blocking activity of vertebrate insulators. *Cell* **98**, 387–96.

70. Burgess-Beusse, B., Farrell, C., Gaszner, M., Litt, M., Mutskov, V., Recillas-Targa, F. *et al.* (2002) The insulation of genes from external enhancers and silencing chromatin. *Proc. Natl. Acad. Sci. USA* **99** (Suppl. 4), 16433–7.

71. Recillas-Targa, F., Pikaart, M. J., Burgess-Beusse, B., Bell, A. C., Litt, M. D., West, A. G. *et al.* (2002) Position-effect protection and enhancer blocking by the chicken β-globin insulator are separable activities. *Proc. Natl. Acad. Sci. USA* **99**, 6883–8.

72. Yusufzai, T. M., Tagami, H., Nakatani, Y., and Felsenfeld, G. (2004) CTCF tethers an insulator to subnuclear sites, suggesting shared insulator mechanisms across species. *Mol. Cell* **13**, 291–8.

73. Kim, T. H., Abdullaev, Z. K., Smith, A. D., Ching, K. A., Loukinov, D. I., Green, R. D. *et al.* (2007) Analysis of the vertebrate insulator protein CTCF-binding sites in the human genome. *Cell* **128**, 1231–45.

74. Splinter, E., Heath, H., Kooren, J., Palstra, R. J., Klous, P., Grosveld, F. *et al.* (2006) CTCF mediates long-range chromatin looping and local histone modification in the β-globin locus. *Genes Dev.* **20**, 2349–54.

75. Parelho, V., Hadjur, S., Spivakov, M., Leleu, M., Sauer, S., Gregson, H. C. *et al.* (2008) Cohesins functionally associate with CTCF on mammalian chromosome arms. *Cell* **132**, 422–33.

76. Wendt, K. S., Yoshida, K., Itoh, T., Bando, M., Koch, B., Schirghuber, E. *et al.* (2008) Cohesin mediates transcriptional insulation by CCCTC-binding factor. *Nature* **451**, 796-801.

77. Yao, S., Osborne, C. S., Bharadwaj, R. R., Pasceri, P., Sukonnik, T., Pannell, D. *et al.* (2003) Retrovirus silencer blocking by the cHS4 insulator is CTCF independent. *Nucleic Acids Res.* **31**, 5317–23.

78. Ishihara, K., Oshimura, M., and Nakao, M. (2006) CTCF-dependent chromatin insulator is linked to epigenetic remodeling. *Mol. Cell* **23**, 733–42.

79. Emery, D. W., Yannaki, E., Tubb, J., and Stamatoyannopoulos, G. (2000) A chromatin insulator protects retrovirus vectors from chromosomal position effects. *Proc. Natl. Acad. Sci. USA* **97**, 9150–5.

80. Rivella, S., Callegari, J. A., May, C., Tan, C. W., and Sadelain, M. (2000) The cHS4 insulator increases the probability of retroviral expression at random chromosomal integration sites. *J. Virol.* **74**, 4679–87.

81. Yannaki, E., Tubb, J., Aker, M., Stamatoyannopoulos, G., and Emery, D. W. (2002) Topological constraints governing the use of the chicken HS4 chromatin insulator in oncoretrovirus vectors. *Mol. Ther.* **5**, 589–98.

82. Aker, M., Tubb, J., Groth, A. C., Bukovsky, A. A., Bell, A. C., Felsenfeld, G. *et al.* (2007) Extended core sequences from the cHS4 insulator are necessary for protecting retroviral vectors from silencing position effects. *Hum. Gene Ther.* **18**, 333–43.

83. Kumar, M., Keller, B., Makalou, N., and Sutton, R. E. (2001) Systematic determination of the packaging limit of lentiviral vectors. *Hum. Gene Ther.* **12**, 1893–1905.

84. Ma, Y., Ramezani, A., Lewis, R., Hawley, R. G., and Thomson, J. A. (2003) High-level sustained transgene expression in human embryonic stem cells using lentiviral vectors. *Stem Cells* **21**, 111–7.

85. Vieyra, D. S. and Goodell, M. A. (2007) Pluripotentiality and conditional transgene regulation in human embryonic stem cells expressing insulated tetracycline-ON transactivator. *Stem Cells* **25**, 2559–66.

86. Kwaks, T. H., Barnett, P., Hemrika, W., Siersma, T., Sewalt, R. G., Satijn, D. P. *et al.* (2003) Identification of anti-repressor elements that confer high and stable protein production in mammalian cells. *Nat. Biotechnol.* **21**, 553–8.

87. Kissler, S., Stern, P., Takahashi, K., Hunter, K., Peterson, L. B., and Wicker, L. S. (2006) *In vivo* RNA interference demonstrates a role for Nramp1 in modifying susceptibility to type 1 diabetes. *Nat. Genet.* **38**, 479–83.

88. Stern, P., Astrof, S., Erkeland, S. J., Schustak, J., Sharp, P. A., and Hynes, R. O. (2008) A system for cre-regulated RNA interference in vivo. *Proc. Natl. Acad. Sci. USA* **105**, 13895–900.

89. Baum, C. and Fehse, B. (2003) Mutagenesis by retroviral transgene insertion: risk assessment and potential alternatives. *Curr. Opin. Mol. Ther.* **5**, 458–62.

90. Nienhuis, A. W., Dunbar, C. E., and Sorrentino, B. P. (2006) Genotoxicity of retroviral integration in hematopoietic cells. *Mol. Ther.* **13**, 1031–49.

91. Ramezani, A., Hawley, T. S., and Hawley, R. G. (2008) Reducing the genotoxic potential of retroviral vectors. *Methods Mol. Biol.* **434**, 183–203.

92. Ramezani, A., Hawley, T. S., and Hawley, R. G. (2008) Combinatorial incorporation of enhancer blocking components of the chicken β-globin 5'HS4 and human T-cell receptor α/δ BEAD-1 insulators in self-inactivating retroviral vectors reduces their genotoxic potential. *Stem Cells* **26**, 3257–66.

93. Modlich, U., Bohne, J., Schmidt, M., Von, K. C., Knoss, S., Schambach, A. *et al.* (2006) Cell-culture assays reveal the importance of retroviral vector design for insertional genotoxicity. *Blood* **108**, 2545–53.

94. Bevis, B. J. and Glick, B. S. (2002) Rapidly maturing variants of the *Discosoma* red fluorescent protein (DsRed). *Nat. Biotechnol.* **20**, 83–7.

95. Ramezani, A. and Hawley, R. G. (2002) Generation of HIV-1-based lentiviral vector particles. *Curr. Protoc. Mol. Biol.* **16.22**, 1–15.

96. Cai, H. and Levine, M. (1995) Modulation of enhancer-promoter interactions by insulators in the *Drosophila* embryo. *Nature* **376**, 533–6.

97. Zhang, F., Thornhill, S. I., Howe, S. J., Ulaganathan, M., Schambach, A., Sinclair, J. *et al.* (2007) Lentiviral vectors containing an enhancer-less ubiquitously acting chromatin opening element (UCOE) provide highly reproducible and stable transgene expression in hematopoietic cells. *Blood* **110**, 1448–57.

98. Buzina, A., Lo, M. Y., Moffett, A., Hotta, A., Fussner, E., Bharadwaj, R. R. *et al.* (2008) β-globin LCR and intron elements cooperate and direct spatial reorganization for gene therapy. *PLoS Genet.* **4**, e1000051.

99. DuBridge, R. B., Tang, P., Hsia, H. C., Leong, P. M., Miller, J. H., and Calos, M. P. (1987) Analysis of mutation in human cells by using an Epstein-Barr virus shuttle system. *Mol. Cell. Biol.* **7**, 379–87.

100. Gorman, C. M., Gies, D., McCray, G., and Huang, M. (1989) The human cytomegalovirus major immediate early promoter can be trans-activated by adenovirus early proteins. *Virology* **171**, 377–85.

101. Hawley, T. S., Herbert, D. J., Eaker, S. S., and Hawley, R. G. (2004) Multiparameter flow cytometry of fluorescent protein reporters. *Methods Mol. Biol.* **263**, 219–38.

102. Gallardo, H. F., Tan, C., Ory, D., and Sadelain, M. (1997) Recombinant retroviruses pseudotyped with the vesicular stomatitis virus G glycoprotein mediate both stable gene transfer and pseudotransduction in human peripheral blood lymphocytes. *Blood* **90**, 952–7.

103. Liu, M. L., Winther, B. L., and Kay, M. A. (1996) Pseudotransduction of hepatocytes by using concentrated pseudotyped vesicular stomatitis virus G glycoprotein (VSV-G)-

Moloney murine leukemia virus-derived retrovirus vectors: comparison of VSV-G and amphotropic vectors for hepatic gene transfer. *J Virol.* **70**, 2497–502.

104. Nightingale, S. J., Hollis, R. P., Pepper, K. A., Petersen, D., Yu, X. J., Yang, C. *et al.* (2006) Transient gene expression by nonintegrating lentiviral vectors. *Mol. Ther.* **13**, 1121–32.

105. Shimotohno, K. and Temin, H. M. (1982) Loss of intervening sequences in genomic mouse α-globin DNA inserted in an infectious retrovirus vector. *Nature* **299**, 265–8.

106. Malim, M. H., Hauber, J., Le, S. Y., Maizel, J. V., and Cullen, B. R. (1989) The HIV-1 rev trans-activator acts through a structured target sequence to activate nuclear export of unspliced viral mRNA. *Nature* **338**, 254–7.

107. Hawley, R. G., Ramezani, A., and Hawley, T. S. (2006) Hematopoietic stem cells. *Methods Enzymol.* **419**, 149–79.

Chapter 6

Integrase Defective, Nonintegrating Lentiviral Vectors

Zuleika Michelini, Donatella Negri, and Andrea Cara

Abstract

Lentiviral vectors are a powerful tool for gene transfer into target cells in vitro and in vivo. However, there are concerns about safety with regard to their use in gene transfer protocols because of insertional mutagenesis following viral infection. Once in the target cells, and in addition to the integrated proviral DNA, lentiviral vectors produce episomal forms of DNA (E-DNA), which are transcriptionally active. Therefore, one strategy to improve safety would envision the block integration of the lentiviral vector while allowing production of E-DNA. Such nonintegrating lentiviral vectors can be produced by introducing mutations in the Integrase (IN) protein of the parental packaging vector. These vectors are fundamentally different from the parental IN competent counterpart, thus opening new avenues for this class of lentiviral vectors as a new gene delivery system for gene therapy strategies, vaccination protocols and as a tool for anti-Integrase drug discovery.

Key words: Integrase, Lentiviral vector, Gene therapy, Vaccine, HIV-1, Episomal DNA

1. Introduction

Lentiviral vectors are very useful tools for gene delivery because of their ability to stably infect a wide variety of cell types, including postmitotic cells in vivo following a single inoculum ((1) and references therein). However, like their wild-type counterpart, lentiviral vectors, such as the parental HIV-1 and SIV, nonspecifically integrate into the target genome (2, 3). As a consequence, safety concerns relative to the use of these vectors for in vivo delivery of therapeutic genes include integration of the vector with consequent risk of insertional mutagenesis (4). To minimize this risk, we and others have devised and engineered Integrase (IN) defective lentiviral vectors to take advantage of the prolonged expression of viral antigens from the circular forms of viral extrachromosomal DNA (E-DNA), thereby giving it a major

Maurizio Federico (ed.), *Lentivirus Gene Engineering Protocols,* Methods in Molecular Biology, vol. 614,
DOI 10.1007/978-1-60761-533-0_6, © Humana Press, a part of Springer Science+Business Media, LLC 2010

safety advantage when compared with the integration competent counterpart ((5) and references therein). During retroviral infection, not only is the proviral DNA integrated into the host genome, but there is also an accumulation of E-DNA, deriving from the circularization of nonintegrated reverse transcribed viral DNA ((6, 7) and references therein). Circular E-DNA is found in infected cells in two different forms containing either a single LTR (1-LTR), or two tandem LTRs (2-LTR). The 1-LTR circles are produced by autointegration reaction, that leads to rearranged circular forms (8, 9) by homologous recombination between the two LTR, or by reverse transcription intermediates failing to complete the reverse transcription process (10); the 2-LTR circles derive from the head to tail joining of the two viral LTR ends of the fully reverse transcribed DNA. Although the integrated DNA is responsible for the production of new viral progeny, the role of E-DNA has remained uncertain. Importantly, we and others have shown that E-DNA is transcriptionally active producing RNA as well as viral proteins, and that circular E-DNA persists in nondividing cells in vivo (5). IN mutant or defective viruses are commonly used to evaluate and characterize the transcriptional activity of E-DNA, and unintegrated lentiviral E-DNA accumulates in nondividing cells if integration is blocked (11). However, the use of IN defective lentiviral vectors has some limitations. In fact, while these forms are long lived in nondividing cells, such as neurons, macrophages and muscle cells, they have short half-lives in cycling cells such as T cells; this is because E-DNA does not have the origin of replications, thus preventing its maintenance in dividing cells. The transient but expression-competent nature of E-DNA has been recently exploited in gene therapy approaches showing targeted genome modifications in cells transduced by IN defective vectors (12, 13). In a different setting, we have shown that the introduction of the SV40 origin of replication allows for the episomal maintenance of the vector which remains transcriptionally active only in cells expressing the corresponding *trans*-acting T antigen (14, 15). On the other hand, the long-lived feature of E-DNA in post mitotic cells has been exploited in immunization protocols in the mouse model (16, 17) and in gene therapy approaches requiring sustained expression in rodent ocular and brain tissues (18) and in muscles. For these reasons, the episomal forms of the vector can be harnessed for effective gene expression in the context of a nonintegrating lentiviral vector for a number of different purposes, depending upon the experimental needs, thus setting the IN defective lentiviral vectors apart from the IN competent counterpart. A list of cell and tissue types that have been successfully transduced with IN defective lentiviral vectors is provided in Table 1. In this paper, we describe how to produce an IN defective lentiviral vector to be used for transduction of cycling cells and nonreplicating macrophages.

Table 1
Cell and tissue types successfully transduced with IN defective lentiviral vectors in vitro and in vivo

	Origin	Cell/tissue type	Expression	Reference
Replicating cells	Human cell line	HeLa, 293, 293 T, C2C12, Jurkat	Transient	(14, 15, 17, 18, 21–26)
	Human cell line	293 T	Stable	(14, 15)
	Feline cell line	CrFK cells	Transient	(11)
	Human primary cells	PBMC, CD34+	Transient	(22, 26)
Growth arrested cells	Human cell line	HeLa, C2C12, HT-1080	Stable	(11, 17, 25)
	Feline cell line	CrFK cells	Stable	(11)
Non replicating cells	Human primary cells	Macrophages, Dendritic cells	Stable	(22, 27)
	Mouse primary cells	Dendritic cells	Stable	(17)
Tissue	Mouse	Muscle, retina, brain	Stable	(18, 24, 25)
	Rat	Neurons, astrocytes, retina	Stable	(11, 21, 28)

2. Materials

2.1. Plasmids and Plasmid Purification

1. Integrase competent packaging plasmid pCMVdR8.2 (14, 16).
2. Integrase defective packaging plasmid pcHelp/IN- (14, 16).
3. Transfer vectors pHR'CMV-GFP or pTY2CMV-GFP (14, 16).
4. Envelope plasmid pHVSV.G (14, 16).
5. HiSpeed Plasmid Maxi Kit with EndoFree Plasmid Buffer Set (Qiagen, Valencia, CA).
6. Endofree plasmid Mega Kit (Qiagen, Valencia, CA).

2.2. Cell Culture and Transfection

1. Human embryonic kidney (HEK) 293 T cell line.
2. Dulbecco's modified Eagle's medium (DMEM, Euroclone, Life Sciences Division, Pero, MI, Italy) supplemented with 10% fetal bovine serum (FBS, Euroclone) and 100 U/ml of penicillin/streptomycin/gentamicine (PSG, Gibco Invitrogen, Paisley, UK).
3. Human primary monocyte-derived macrophages.

4. Monocyte/macrophage medium: Roswell Park Memorial Institute (RPMI) 1640 medium (Euroclone) supplemented with 10% fetal bovine serum (FBS, Euroclone), 100 U/ml of PSG, and 40 U/ml of Granulocyte macrophage colony-stimulating factor (GM-CSF, R&D systems, Minneapolis, MN).

5. Solution of trypsin (0.25%) and ethylenediaminetetraacetic acid (EDTA) 1 mM (Gibco/Invitrogen).

6. Calcium Phosphate-based Profection Mammalian Transfection System (Promega Corporation, Madison WI).

2.3. Virus Concentration

1. Filters with 0.45 µM pore size (Millipore Corporation, Billerica, MA).

2. Filter unit with 0.45 µM pore size (Millipore).

3. 20% Sucrose solution in 1× Phosphate buffered saline (PBS).

2.4. Viral Titer Evaluation

1. HIV-1 CAp24 enzyme-linked immunosorbent assays (Innogenetics, Ghent, Belgium).

2. Reverse Transcriptase (RT) assay reagents (19). MIX for 10 samples is as follows:

Tris-Cl 1 M pH 7.4	40 µl
DTT 0.02 M	200 µl
$MgCl_2$ 0.5 M	20 µl
KCl 1 M	45 µl
PolyA/Oligo dT	50 µl
^{3}HTTP	20 µl
ddH_2O	325 µl

3. Fluorescence microscope.

4. FACSCalibur analytical flow cytometer with CellQuest software (BD Biosciences Immunocytometry Systems, San Jose, CA).

3. Methods

Primary cells and live animals are sensitive to lipopolysaccharide (LPS), an endotoxin found on the outer membrane of Gram-negative bacteria. Consequently, if recombinant lentiviral particles are to be used in vivo (i.e. mouse model) or on primary cells such as macrophages (MΦ), preparation of plasmids should be performed using endotoxin-free buffers. This allows for almost complete elimination of bacterial endotoxin carryover contaminants during plasmid DNA Maxi/Megapreps. Plasmids used for transfections include a transfer vector expressing the model antigen green fluorescent protein (GFP) (pHR'CMV-GFP or pTY2CMV-GFP), a packaging plasmid IN competent (pCMVdR8.2) or a packaging plasmid IN defective (pcHelp/IN-), and an envelope expressing plasmid (pHVSV.G). Plasmids should be transfected with a 3:2:1 ratio, respectively, for the Transfer:Packaging:Envelope plasmids. This ensures optimal recombinant viral particles production. The choice of the packaging plasmid depends on the experimental conditions. We generally use packaging plasmids expressing all viral proteins with the exclusion of envelope (provided in *trans* by the pHVSV.G plasmid) in the IN competent packaging plasmid and of both envelope and the IN coding sequence in the IN defective packaging plasmid. This is because some cell types may need the presence of accessory genes (such as *vif*, *vpr* or *nef*) for optimal transduction levels (20). In the case this is not necessary, minimal packaging plasmids devoid of viral accessory genes may be used instead (see Note 1).

3.1. Transfection

Transfection must be performed following standard methodologies as indicated in any protocol for lentiviral vector production. Below is a brief outline of the method using 293 T cells. Cells are usually plated on Monday so that all steps can be easily completed during the week.

Day 1: *Plating* (4–6 p.m.).

Plate $3–3.5 \times 10^6$ of 293 T cells per 10 cm Petri dish in 10 ml complete D-MEM (next morning you should have around 80% confluency).

Day 2: *Transfection* (9–10 a.m.)

1. Change the medium (9 ml of complete D-MEM),

2. Prepare the calcium-phosphate precipitate (0.5 ml/10 cm plate). We use the Profection Mammalian Calcium Phosphate transfection kit from Promega in either the 10 ml snap-cap or the classical 15 ml Falcon polypropilene tubes (never use polystirene tubes as the DNA tends to stick to the walls of the tube). First, add the water (as specified in the manufacturer's instructions) then each of the following plasmids in the amounts indicated below:

   • Transfer vector (pHR'CMV-GFP or pTY2CMV-GFP plasmids): 15 µg

   • Packaging plasmid (CMVΔR8.2 (IN competent) or pcHelp/IN-(IN defective)): 10 µg

   • Envelope plasmid (HVSV.G): 5 µg.

3. Add 62 µl 2.5 M $CaCl_2$. Add this solution to 0.5 ml 2× HBS by gently vortexing following the manufacturer's instructions as reported in the kit. Keep at room temperature for 25–30 min, then add dropwise to each plate and swirl gently the plate.

4. Change medium (6–8 h later); remove medium with precipitate and replace with 7 ml of fresh D-MEM for each plate.

If you are planning to prepare a large stock of virus, you should prepare 5–10 plates for each virus (see Note 2).

Day 4: *Collection* of supernatants containing recombinant virus (10–12 a.m.)

1. Collect the viral supernatant containing medium in 15-ml Falcon tube.

2. Spin at $700 \times g/10$ min/$4°C$.

3. Filter through $0.45\,\mu m$ pore size filters (it is recommended to never use the $0.22\,\mu m$ pore size filters otherwise viral titers will decrease).

At this point the virus can be used for transduction, frozen at $-70°C$ for future use or concentrated.

3.2. Lentiviral Vector Concentration

1. For transduction of primary cells or for use in vivo, the virus should be concentrated on a sucrose cushion.

2. Put 6 ml of 20% sucrose (prepared in TNE, filtered through a $0.22\,\mu m$ pore size filter and kept at $4°C$) on the bottom of a 33-ml Beckman conical tube and overlay slowly with 26 ml of viral supernatant. The ultracentrifugation through the sucrose gradient allows for almost complete elimination of DNA carryover and protein contaminants which may interfere with transduction. Subsequently spin at $50,000 \times g/2h/4°C$ in a Beckman SW28 swinging bucket rotor.

3. After spin, discard supernatant, put the tubes on ice and resuspend the virus in a desired volume (100–200 µl) of PBS in 1% bovine serum albumin (BSA, but avoid BSA in immunization protocols).

4. Aliquot and store at $-70°C$ or liquid Nitrogen.

3.3. Lentiviral Vector Titer Evaluation

Viral titers, but most importantly efficiency of transfection and viral recovery, are normalized by measuring the HIV-1 CAp24 content (enzyme-linked immunosorbent assays, Innogenetics, Ghent, Belgium), and the RT activity (19). For evaluation of RT activity:

1. Centrifuge 0.5–1 ml of virus containing supernatant in a Beckman TL-100 ($80,000 \times g/15$ min/$4°C$).

2. Discard the supernatant (tap the tube on adsorbent paper to remove any liquid leftover).

3. Resuspend pellet (which sometimes is not visible) in 100 µl of TNE/Tritox X-100 0.1%.

4. Put on ice for 20 min.

5. After incubation, put 30 µl of each sample in a 5 ml snap polystyrene cap tube, and add 70 µl of MIX (*see* below). For the negative control, you will have 30 µl TNE/Tritox X-100 0.1%. plus 70 µl of MIX.

6. Vortex every time you add a component to the sample.

7. Put samples 1 h at 37°C for the RT reaction to develop.

8. Stop reaction by addition of 500 µl Sodium Pyrophosphate 0.01 M and 600 µl of TCA 20%.

9. Leave on ice for 20 min.

10. Filter on nitrocellulose filters (0.45 µm, Sartorius) by washing with TCA 5%/SDS 0.1% and 2 more washes with TCA 5%.

11. Put filters (if dry wet them with TCA 5%) in vials with 5 ml scintillation fluid and read counts in a beta counter. Read again the next day as it is more accurate.

3.4. Transduction of 293 and 293 T Cells

Normalized amounts of viral preparation are used to transduce target 293 T cells. Viral preparations are always tested on either of these cell types to evaluate the efficiency of recombinant virus recovery (an example is shown in Fig. 1). This provides an indication of the quality of the preparation that are subsequently to be used in other systems (primary cells or in vivo).

3.5. Macrophage Preparation from Primary Human Monocytes

1. Pre-treat flasks/wells with a layer of human AB serum for 30 min at R.T. Discharge the serum.

2. To seed macrophages, follow the following proportions:

    – for 96 well plates:1.25×10^6 PBMC in 0.25 ml medium/well,

    – for 48 well plates: 2.5×10^6 PBMC in 1 ml medium/well,

    – for 24 well plates: 5×10^6 PBMC in 1 ml medium/well,

    – for 12 well plates: 10^7 PBMC in 2 ml medium/well,

    – for 6 well plates: 2×10^7 PBMC in 3 ml medium/well,

    – for 25 cm² flask : 5×10^7 PBMC in 5 ml medium/flask.

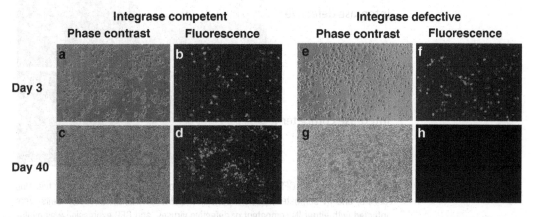

Fig. 1. In vitro transduction of 293 cells with IN competent and IN defective HR'CMV-GFP vector at early (3 days, *panels* (**a**), (**b**), (**e**) and (**f**)) and late (40 days, *panels* (**c**), (**d**), (**g**) and (**h**)) time points. (**a**)–(**d**), transduction with IN competent. (**e**)–(**h**), transduction with IN defective. Both days after transduction and phase contrast/fluorescence images are indicated. Cells were split periodically during the course of experiment. Expression from E-DNA from the IN defective vector is visible only at early time points as E-DNA is lost following 293 cellular division. As expected, expression from the IN competent vector is visible at all time points

3. Check the colour of the medium. If it does not turn yellow before, change the medium after three days by eliminating at the same time the cells that are not adherent.

4. Check the cell cultures from time to time at the microscope. They should be 50–90% confluent, forming numerous colonies, most of which are visible also without microscope.

5. If necessary, change the medium. Medium should be changed independently of its color 1 week after seeding.

3.6. Macrophage Transduction

1. Transduce macrophages with the concentrated or simply clarified vector preparations not before 10 days after seeding (range 10–14 days).

2. One day before the transduction, change the medium and carefully remove all the residual nonadherent cells.

3. Transduce the cells using a minimum volume of recombinant vector (e.g. 300 μl/well or 2 ml/flask). Incubate for 4 h in the CO_2 incubator, then add medium up to 1 ml/well or 5 ml/flask. Incubate overnight (see Note 3).

4. The day after, wash 3 times with warm PBS.

5. Add 1–3 ml/well or 5 ml/flask of fresh medium.

6. Harvest at the desired time points (see Note 4).

7. The outcome from a typical experiment of macrophage transduction is given in Fig. 2 (see Note 5).

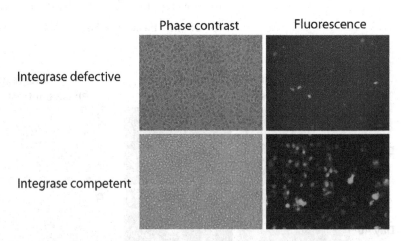

Fig. 2. Evaluation of GFP expression in macrophages infected with IN defective (*top panels*) and IN competent (*bottom panels*) TY2CMV-GFP lentiviral vector. Cells were infected with either IN competent or defective viruses, and GFP expression was evaluated at 14 day by both light and fluorescent microscopy. Expression from E-DNA transduced with the IN defective vector is visible at late time point as E-DNA is maintained in nondividing macrophages. As expected, expression from the IN defective vector is lower than that from the IN competent counterpart

4. Notes

1. The www.addgene.org website contains and sells a number of different packaging and envelope expressing plasmids in addition to several transfer vector plasmid expressing different transgenes.

2. We prefer to perform separately each transfection and not in batches. However, a single master mix solution can be prepared containing all plasmid DNA and the Calcium Phosphate. This solution can be dispensed into each tube corresponding to each transfection containing the 2× HBS solution.

3. Use at least 1×10^5 cpm of RT and normalize on the basis of RT values in case more than one vector is tested.

4. Usually, the transcription of the transgene under these conditions starts the day after the lentiviral vector challenge. A reasonable time course could be to harvest at 1, 5, 10, 15, 20 and eventually 25 through 60 days after transduction.

5. The transduction efficiency is extremely variable (donor dependent). In addition, transcription from unintegrated templates is much lower than that from the integrated counterpart, as indicated by the lower fluorescence intensity detectable after transduction.

Acknowledgments

The authors would like to thank Patrizia Cocco and Ferdinando Costa (National AIDS Center, Istituto Superiore di Sanità, Rome, Italy) for technical support.

References

1. Wiznerowicz, M., and Trono, D. (2005) Harnessing HIV for therapy, basic research and biotechnology. *Trends Biotechnol.* **23**, 42–7.

2. Hematti, P., Hong, B. K., Ferguson, C., Adler, R., Hanawa, H., Sellers, S., Holt, I. E., Eckfeldt, C. E., Sharma, Y., Schmidt, M., von Kalle, C., Persons, D.A., Billings, E. M., Verfaillie, C. M., Nienhuis, A. W., Wolfsberg, T. G., Dunbar, C. E., and Calmels, B. (2004) Distinct genomic integration of MLV and SIV vectors in primate hematopoietic stem and progenitor cells. *PLoS Biol.* **2**, e423.

3. Mitchell, R. S., Beitzel, B. F., Schroder, A. R., Shinn, P., Chen, H., Berry, C. C., Ecker, J. R., and Bushman, F. D. (2004) Retroviral DNA integration: ASLV, HIV, and MLV show distinct target site preferences. *PLoS Biol.* **2**, e234.

4. Hacein-Bey-Abina, S., von Kalle, C., Schmidt, M., Le Deist, F., Wulffraat, N., McIntyre, E., Radford, I., Villeval, J. L., Fraser, C. C., Cavazzana-Calvo, M., and Fischer, A. (2003) A serious adverse event after successful gene therapy for X-linked severe combined immunodeficiency. *N. Engl. J. Med.* **348**, 255–6.

5. Cara, A., and Klotman, M. E. (2006) Retroviral E-DNA: persistence and gene expression in nondividing immune cells. *J. Leukoc. Biol.* **80**, 1013–17.

6. Cara, A., Vargas, J. Jr., Keller, M., Jones, S., Mosoian, A., Gurtman, A., Cohen, A., Parkas, V.,

Wallach, F., Chusid, E., Gelman, I.H., and Klotman, M.E. (2002) Circular viral DNA and anomalous junction sequence in PBMC of HIV-infected individuals with no detectable plasma HIV RNA. *Virology.* **292**, 1–5.

7. Cara, A., and Reitz, M. S., Jr. (1997) New insight on the role of extrachromosomal retroviral DNA. *Leukemia* **11**, 1395–99.

8. Shoemaker, C., Goff, S., Gilboa, E., Paskind, M., Mitra, S. W., and Baltimore, D. (1980) Structure of a cloned circular Moloney murine leukemia virus DNA molecule containing an inverted segment: implications for retrovirus integration. *Proc. Natl. Acad. Sci. USA* **77**, 3932–36.

9. Farnet, C. M., and Haseltine, W. A. (1991) Circularization of human immunodeficiency virus type 1 DNA in vitro. *J. Virol.* **65**, 6942–52.

10. Miller, M. D., Wang, B., and Bushman, F. D. (1995) Human immunodeficiency virus type 1 preintegration complexes containing discontinuous plus strands are competent to integrate in vitro. *J. Virol.* **69**, 3938–44.

11. Saenz, D. T., Loewen, N., Peretz, M., Whitwam, T., Barraza, R., Howell, K. G., Holmes, J. M., Good, M., and Poeschla, E. M. (2004) Unintegrated lentivirus DNA persistence and accessibility to expression in nondividing cells: analysis with class I integrase mutants. *J. Virol.* **78**, 2906–20.

12. Cornu, T. I., and Cathomen, T. (2007) Targeted genome modifications using integrase-deficient lentiviral vectors. *Mol. Ther.* **15**, 2107–13.

13. Lombardo, A., Genovese, P., Beausejour, C. M., Colleoni, S., Lee, Y. L., Kim, K. A., Ando, D., Urnov, F. D., Galli, C., Gregory, P. D., Holmes, M. C., and Naldini, L. (2007) Gene editing in human stem cells using zinc finger nucleases and integrase-defective lentiviral vector delivery. *Nat. Biotechnol.* **25**, 1298–306.

14. Vargas, J. Jr., Gusella, G. L., Najfeld, V., Klotman, M. E., and Cara, A. (2004) Novel integrase-defective lentiviral episomal vectors for gene transfer. *Hum. Gene Ther.* **15**, 361–72.

15. Vargas, J. Jr., Klotman, M.E., and Cara, A. (2008) Conditionally replicating lentiviral-hybrid episomal vectors for suicide gene therapy. *Antiviral Res.* **80**, 288–94.

16. Negri, D. R., Michelini, Z., Baroncelli, S., Spada, M., Vendetti, S., Buffa, V., Bona, R., Leone, P., Klotman, M. E., and Cara, A. (2007) Successful immunization with a single injection of non-integrating lentiviral vector. *Mol. Ther.* **15**, 1716–23.

17. Coutant, F., Frenkiel, M.P., Despres, P., and Charneau, P. (2008) Protective antiviral immunity conferred by a nonintegrative lentiviral vector-based vaccine. *PLoS ONE.* **3**, e3973.

18. Yáñez-Muñoz, R. J., Balaggan, K. S., MacNeil, A., Howe, S. J., Schmidt, M., Smith, A. J., Buch, P., MacLaren, R. E., Anderson, P. N., Barker, S. E., Duran, Y., Bartholomae, C., von Kalle, C., Heckenlively, J. R., Kinnon, C., Ali, R. R., and Thrasher, A. J. (2006) Effective gene therapy with nonintegrating lentiviral vectors. *Nat. Med.* **12**, 348–53.

19. Goff, S., Traktman, P., and Baltimore, D. (1981) Isolation and properties of Moloney murine leukemia virus mutants: use of a rapid assay for release of virion reverse transcriptase. *J. Virol.* **38**, 239–48.

20. Zufferey, R., Nagy, D., Mandel, R. J., Naldini, L., Trono, D. (1987) Multiply attenuated lentiviral vector achieves efficient gene delivery in vivo. *Nat. Biotechnol.* **15**, 871–75.

21. Bayer, M., Kantor, B., Cockrell, A., Ma, H., Zeithaml, B., Li, X., McCown, T., and Kafri, T. (2008) A large U3 deletion causes increased in vivo expression from a nonintegrating lentiviral vector. *Mol. Ther.* **16**, 1968–76.

22. Berger, G., Goujon, C., Darlix, J.L., and Cimarelli, A. (2009) SIVMAC Vpx improves the transduction of dendritic cells with nonintegrative HIV-1-derived vectors. *Gene Ther.* **16**, 159–63.

23. Philpott, N.J., and Thrasher, A.J. (2007) Use of nonintegrating lentiviral vectors for gene therapy. *Hum. Gene Ther.* **18**, 483–9.

24. Philippe, S., Sarkis, C., Barkats, M., Mammeri, H., Ladroue, C., Petit, C., Mallet, J., and Serguera, C. (2006) Lentiviral vectors with a defective integrase allow efficient and sustained transgene expression in vitro and in vivo. *Proc. Natl. Acad. Sci. USA.* **103**, 17684–9.

25. Apolonia, L., Waddington, S.N., Fernandes, C., Ward, N.J., Bouma, G., Blundell, M.P., Thrasher, A.J., Collins, M.K., and Philpott, N.J. (2007) Stable gene transfer to muscle using non-integrating lentiviral vectors. *Mol. Ther.* **15**, 1947–54.

26. Nightingale, S.J., Hollis, R.P., Pepper, K.A., Petersen, D., Yu, X.J., Yang, C., Bahner, I., and Kohn, D.B. (2006) Transient gene expression by nonintegrating lentiviral vectors. *Mol. Ther.* **13**, 1121–32.

27. Gillim-Ross, L., Cara, A., and Klotman, M.E. (2005) HIV-1 extrachromosomal 2-LTR circular DNA is long-lived in human macrophages. *Viral Immunol.* **18**, 190–6.

28. Rahim, A.A., Wong, A.M., Howe, S.J., Buckley, S.M., Acosta-Saltos, A.D., Elston, K.E., Ward, N.J., Philpott, N.J., Cooper, J.D., Anderson, P.N., Waddington, S.N., Thrasher, A.J., and Raivich, G. (2009) Efficient gene delivery to the adult and fetal CNS using pseudotyped non-integrating lentiviral vectors. *Gene Ther.* Jan 22. doi:10.1038/gt.2008.186.

Chapter 7

Lentivirus-Based Virus-Like Particles as a New Protein Delivery Tool

Claudia Muratori, Roberta Bona, and Maurizio Federico

Abstract

Virus Like Particles (VLPs) are self-assembling, nonreplicating, nonpathogenic, genomeless particles similar in size and conformation to intact infectious virions. The possibility of engineering VLPs to incorporate heterologous polypeptides/proteins renders VLPs attractive candidates for vaccine strategies, as well as for protein delivery for basic science. Among the wide number of VLP types, our expertise focused on both retro- and lentivirus based VLPs as protein delivery tools. In particular, here we describe a system relying on the finding that some HIV-1 Nef mutants are incorporated at high levels into both Human Immunodeficiency virus (HIV)-1 and Moloney Leukemia Virus (MLV)-based VLPs. Most importantly, these Nef mutants can efficiently act as anchoring proteins upon fusion with heterologous proteins up to 630 amino acids in length. This chapter describes the preparation of prototypic HIV-1 based VLPs incorporating Nef mutant-GFP fusion molecules. Besides having potential utility in the field of basic virology, these VLPs represent a useful reference model for recovering alternative retro- or lentiviral based VLPs for the cell delivery of polypeptides/proteins of interest.

Key words: Virus-like particles, Protein delivery, Nef, GFP, Virion incorporation, Fusion proteins, FACS analysis

1. Introduction

Virus-like particles (VLPs) are genomeless virions produced by the assembling of one or more viral structural proteins (1). They can be generated by engineering the genomes of many virus species, including human papillomavirus (HPV) (2, 3), human hepatitis B and C viruses (HBV, HCV) (4, 5), and retro- (6) and lentiviruses (for a review, see ref. 7). VLPs can be produced from different mammalian cell lines, various species of yeast (i.e. *Saccharomyces cerevisiae* or *Pichia pastoris*) (8, 9), *Escherichia coli* (10) or other bacterial systems, and the baculovirus/insect cell

Maurizio Federico (ed.), *Lentivirus Gene Engineering Protocols*, Methods in Molecular Biology, vol. 614,
DOI 10.1007/978-1-60761-533-0_7, © Humana Press, a part of Springer Science+Business Media, LLC 2010

system (11, 12). Ease of expression, ability to scale-up, and cost of production have made yeast the most popular VLP expression system. However, considerations such as appropriate protein glycosylation, correct folding, and assembly favor alternative methods of production. Among these, mammalian cells produce appropriate modifications and authentic assembly, being, however, a less controllable and more expensive expression system. In the case of production of VLPs for immunization studies, the choice of the expression system for VLP production might significantly influence the direction and outcome of the immune response. In any case, VLPs possess excellent adjuvant properties capable of inducing both cellular and humoral immune responses (1, 13). Commercialized VLP-based vaccines have been successful in protecting humans from HBV (14) and HPV infections (15), and are currently being explored for their potential to challenge other infectious diseases and cancer.

However, the use of VLPs in vaccinology is just one of many potential applications within the VLP field. The ability of VLPs to include heterologous nucleic acids and small molecules has made them novel vessels for gene and drug delivery. HPV VLPs have been shown to mediate functional delivery of plasmid DNA in vivo (16). MS2 bacteriophage VLPs have been used to deliver antisense oligodeoxynucleotides (ODNs) (17) suggesting that RNA phage may be useful as a new nanomaterial for delivering antisense ODNs or other drug candidates. Lately, VLPs have also been used as scaffolds in nanoparticle biothecnology. In fact, Wang et al. had previously exploited the possibility of chemically modifying mutant cowpea mosaic virus particles by exposing sulfhydryl groups in order to attach fluorescent dyes and gold clusters (18). Likewise, the tobacco mosaic virus-derived capsid protein has proven to be a suitable scaffold for extensive chemical modification allowing assembly of nanobiopolymers (19). Yet another study has described the usage of recombinant M13 phage particles as organic templates to polymerize nanowires as building blocks for semiconductors or magnetic materials (20).

Among different VLP types, HIV-1 derived VLPs are prime candidates as carriers for the delivery of foreign amino acid sequences. Two different strategies have been applied to incorporate polypeptides into these VLPs. In type I VLPs, small epitopes are integrated or fused with the Gag polyprotein, which directs VLP formation (21). In type II VLPs, the foreign protein is incorporated into the envelope at the outer particle surface (22). However, the major disadvantage of both VLP types is the limitation of maximal acceptable size of inserted amino acid sequences. For instance, VLPs incorporating foreign polypeptides fused in frame with Gag assemble with efficiency that strongly diminishes with the length of the heterologous sequences (23, 24). In an effort to circumvent these limitations, the insertion of foreign sequences in auxiliary/regulatory proteins incorporating in virions at valuable extents (i.e., Vpr, Vpx and Nef) (25, 26) has been successfully attempted. In particular, Vpr has been reported to incorporate at basically unchanged levels when either CAT (27), GFP (28) or β-lactamase (29) are fused at its N-terminus. Similarly, we recently found that the G3C amino acid substitution (30) (creating a palmitoylation site at the N-terminus), or the coexisting V153L and E177G Nef mutations (26) (and also the combination of the three mutations, unpublished results) lead to surprisingly high incorporation in both HIV-1 and MLV virions or VLPs, i.e., about 100-fold higher than the wt counterpart. Concerning HIV-1, whatever Nef mutant (Nef^{mut}) considered, the mechanism is thought to be the increase of its localization at cell membrane rafts (26, 31), that are the sites from where HIV-1 preferentially buds (32).

We already exploited the unexpected phenotype of these Nef^{mut} in different ways. In particular, we recovered fluorescent HIV-1 VLPs through VLP incorporation of GFP fused at the C-terminus of Nef G3C (30). In addition, we optimized an unconventional method for the specific elimination of HIV-1 infected cells by the delivery of V153L/E177G Nef-HSV/TK fusion products through HIV-1 based VLPs pseudotyped with CD4 and CXCR4 or CCR5 receptors (33). Considering that Nef^{mut} can be fused with heterologous proteins up to 630 amino acids without relevant loss of incorporation efficiency (unpublished results), Nef^{mut}-based VLPs can be considered an excellent tool of protein delivery for many purposes, e.g., immunization studies, analysis of virus-cell interaction, evaluation of the activity of antiviral compounds. Notably, the V153L E177G mutated Nef has been successfully used also by Green and coll. for enhancing the viral incorporation of the cellular anti-HIV-1 protein APOBEC3G (34).

We describe here the recovery of VSV-G pseudotyped, fluorescent HIV-1 VLPs in great detail. These VLPs have been found entering different types of antigen presenting cells with high efficiency (Fig. 1). These methods can be easily translated for the production of VLPs incorporating virtually any kind of foreign amino acid sequence.

2. Materials

2.1. Recovery of the Molecular Constructs

1. Plasmid carrying the sequence of a wt HIV-1 isoform or, alternatively, a wt Nef allele (see Note 1).

2. Plasmid carrying the sequences of the gene (polypeptide) of interest (see Note 2).

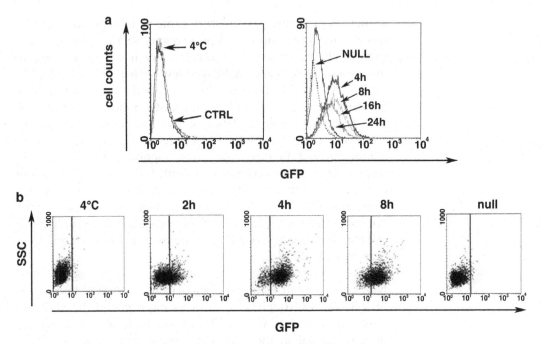

Fig. 1. Kinetics of internalization of (VSV-G)Nef^{mut}-GFP VLPs in antigen presenting cells. (**a**) Left panel: human B lympho-blastoid cells (B-LCLs) were untreated or challenged with (VSV-G)Nef^{mut}-GFP VLPs, and maintained for 4 h at 4°C, and finally FACS analyzed. Right panel: human B-LCLs were challenged with (VSV-G) Nef^{mut}-GFP VLPs or with the receptor-less counterpart (Null), and FACS analyzed at different times. Cells incubated with Null VLPs were analyzed only at 4 hours after the challenge. (**b**) Kinetic of internalization of (VSV-G) Nef^{mut}-GFP VLPs in immature DCs. Shown are the FACS analyses carried out on immature DCs at different times after the challenge with either (VSV-G) Nef^{mut}-GFP VLPs, or 4 h after the challenge with the receptor-less counterpart (Null VLPs). As control, (VSV-G) Nef^{mut}-GFP VLP treated cells were also maintained for 4 h at 4°C. In all cases, samples were treated for 10 min with trypsin before FACS analysis

3. An immediate-early CMV promoter regulated eukaryotic vector where the sequences of the Nef^{mut}-based fusion protein can be accommodated (see Note 3).

4. Eukaryotic vector expressing the G protein from Vesicular Stomatitis virus (VSV-G).

5. Optional: eukaryotic vector expressing a HIV-1 packaging construct, e.g., pCMVΔR8.74 (see Note 4).

2.2. VLP Producing Cell System

1. Dulbecco's Modified Eagle's Medium (DMEM) (Gibco-BRL, Bethesda, MD). Supplemented with 10% fetal bovine serum (FBS, Gibco), heat inactivated at 56°C for 30 min.

2. HIV-1 packaging 293/ Rev-Gag-Pol (here referred to as 293/GPR) cells (35) (see Note 5).

3. Sodium Butyrate (Sigma-Aldrich, St. Louis, MO). Dissolve at 0.5–1 M concentration in 1× Phosphate Buffer Saline (PBS), sterilize by 0.45 μM pore diameter filter, aliquot, and store at −20°C.

4. Ponasterone A (PonA) (Sigma-Aldrich). Dissolve in DMSO at 1 mM.

5. Lipofectamine 2000 (Invitrogen, Carlsbad, CA).

2.3. VLP Concentration, Purification, and Titration

1. Polyallomer ultracentrifuge tubes (see Note 6).

2. 20% sucrose solution in 0.9% sodium chloride.

3. Quantitative anti-HIV-1 CAp24 mAb-based ELISA (Innogenetics, Gent, Belgium) (see Note 7).

4. HIV-1 reverse transcriptase reaction mix: 40 mM Tris-HCl pH 7.4, 4 mM DTT, 45 mM KCl, 10 mM $MgCl_2$, 50 µg/ml poly (rA), 5 µg/ml oligo(dT)$_{12-18}$, 2 µCi ^{3}H TTP (see Note 8).

2.4. VLP Characterization: Western Blot Analysis

1. Equipments and buffers for casting 10–12% SDS-polyacrilamide gel electrophoresis (PAGE).

2. Subtilisin A (Sigma-Aldrich). Dissolve in 3D H_2O at 10 mg/ml, aliquot, and store at –80°C. 2× subtilisin buffer: 40 mM Tris-HCl (pH 8.0), 2 mM $CaCl_2$ 200 mM NaCl Sterilize by filtration, aliquot, and store at –20°C.

3. Recombinant (r)Nef protein. This can be obtained from NIBSC AIDS Reagent Program, or purchased from either EBI, (Frederick, MD), or Intracel (Issaquah, WA).

4. Equipments and buffers for Western blot analysis.

5. Primary antibodies for Western blot analysis: for Gag detection, pool of strongly HIV-1 positive human sera (see Note 9); anti-HIV-1 Nef ARP 444 from NISBC AIDS Reagent Program (see Note 10); anti-VSV-G from Immunology Consultant Laboratories (Newberg, OR). (see Note 11); anti-GFP, from Clontech (Palo Alto, CA) (see Note 12).

6. Secondary Abs for Western blot analysis: horseradish peroxidase (HRP)-conjugated anti-rabbit/mouse/sheep/human Abs from Pierce (Rockford, IL).

7. Chemioluminescent detection reagents (see Note 13).

2.5. VLP Characterization: FACS Analysis

1. Surfactant-free white aldehyde/sulfate latex beads (Invitrogen Molecular Probes, Eugene, ON).

2. Kit for cell permeabilization Cytofix/Cytoperm (BD Bioscience).

3. MAbs for detecting VLP-associated proteins: PE-conjugated anti-CAp24 KC-57 (Coulter Corp. Hialeah, FL), anti-Nef mAb clone 6.2 from NIH AIDS Research and Reference Reagent Program. Abs against envelope proteins and the heterologous moiety of choice fused with Nef^{mut} working well in FACS analysis (see Note 14).

4. BD FACS Calibur cytofluorimeter running Cell Quest software, or equivalent.

3. Methods

The first and, perhaps, most time consuming step is the design and the production of vectors expressing Nef^mut-based fusion protein. Conversely, vectors expressing the envelope protein of choice and, optionally, the HIV-1 packaging construct are largely popular reagents, thus being easy to be obtained or purchased. We observed good VLP incorporation with either single G3C amino acid substitution, double V153L and G177E mutations, or combining the three Nef mutations. Of course, for the sake of simplicity, the single G3C mutation is generally preferably but, in case the numerous anti-cellular effects of wt Nef (36) are expected to be detrimental for the overall experimental design, inclusion of the two additional mutations are recommended. In fact, we and others provide evidence that the V153L/E177G Nef mutant lacks the most part of the effects typically induced by the wt counterpart, i.e., MHC Class I and CD4 down-regulation, PAK-2 activation, inhibition of vacuolar ATPase, and increase of HIV infectivity (34, 37).

Here, we describe the methods for production, concentration, and titration of HIV-1 based VLPs incorporating the NefG3C-GFP fusion product. The vector expressing this fusion protein was obtained from the laboratory of O. Schwartz, Institute Pasteur, France, and is freely available upon request. Critical points for the construction of vectors expressing alternative, NefG3C-based fusion are also reported. For the inclusion of the additional Nef mutations, site-specific Nef mutagenesis should be performed. Alternatively, the vector expressing the V153L and G177E double mutant is available from our laboratory.

3.1. Recovery of Vectors Expressing the Nef^mut-based Fusion Protein

In the case of the recovery of a vector expressing an Nef^mut-based fusion protein other than NefG3C-GFP is planned, amplify wt Nef and the polypeptide/protein sequence of interest separately by conventional PCR procedures. The final sequence codifying the fusion protein will be recovered by overlapping PCR procedures. In this regard, considering that the Nef moiety must be placed at the N-terminal of the fusion protein, design appropriate overlapping 3' Nef and 5' heterologous sequence primers. These primers should include a complementary sequence extending at least 10–15 nucleotides. Pay attention to insert the G3C mutation in the 5' Nef primer. This can be done simply by inserting a single nucleotide mutation in the third Nef triplet, i.e. ATG GGT TGC AAG TGG TCA, where the mismatched nucleotide is underlined. On the basis of experience we acquired with more than ten Nef^mut based fusion proteins we successfully constructed, no spacer amino acid sequences are needed between Nef^mut and the heterologous polypeptide/protein. Finally, insert appropriate restriction enzyme sequences in the 5' Nef primer as well as in the 3' primer of the polypeptide/protein to be fused with Nef.

3.2. VLP Producing Cell System

1. For 15 cm diameter dishes, seed 2×10^7 293/GPR cells in DMEM 10% FBS without antibiotics the day before transfection (see Note 15).

2. The day of transfection, bring the volume of the medium to 12 ml. Add 3 ml of mix of transfection (i.e., Lipofectamine 2000 plus DNA) according to manufacturer's recommendations.

3. Eight hours later, add sodium butyrate and ponasterone A at final concentrations of 5 mM and 1 μM, respectively (see Note 16).

4. The day after, repeat the induction procedure, however replacing the medium. At this time, antibiotics may be included. Pay special attention to leave the packaging cells attached.

5. The day after, harvest the supernatants. Additional supernatant harvestings may be done every 8–16 h for 2–3 days, according to the cell viability.

6. Clarify the supernatants either by centrifugation ($1,500 \times g$ for 30 min at 4°C) or by filtration through 0.45 μM pore diameter filters. At this time, supernatants can be stored at –80°C.

3.3. VLP Concentration, Purification, and Titration

1. To concentrate and partially purify the VLP contained in clarified supernatants, load polyallomer SW 28 tubes with 6 ml of 20% sucrose solution, and then stratify the supernatants very slowly. Volumes up to 30 ml can be easily accommodated. To avoid tube collapsing, they should be loaded with at least 30 ml of total volume. Ultracentrifuge 2.5 h at $50,000 \times g$, 4°C.

2. Decant the supernatants by inverting the tubes, eliminate possible residual drops with sterile paper, and add 100–200 μL of medium without serum to the VLP pellet (that is generally not visible).

3. Let stand the tubes at 4°C for at least 4 h, taking care to preserve sterility, e.g., by sealing them with Parafilm (see Note 17).

4. By gently scraping the bottom of the tube with the tip, harvest the resuspension volume, aliquot, and store at –80°C. Make at least two 1:100 diluted aliquots for titration. Dilute 4 μl of the 1:100 aliquot in 100 μl of TNE-Triton X-100 0.1% for reverse transcriptase assay.

5. For ELISA titration, make $1:10^3$ to $1:10^6$ dilutions for each VLP preparation. Follow the protocol specific for the kit ELISA of choice. For 15 cm diameter plate, recovery of 20–100 μg of CAp24 for each harvesting could be considered an acceptable outcome.

6. For the reverse transcriptase assay, we routinely perform an in house made test. This is performed in triplicate by incubating a total of 30 μl of VLP samples diluted as described earlier, with 70 μl of the RT reaction mix. After 1 h of incubation at 37°C, the reaction is stopped with 0.5 ml of 0.1 M sodium pyrophosphate, pH 5, and radiolabeled nucleic acids precipitated

by adding 0.6 ml of 20% trichloroacetic acid. After 1 h of incubation on ice, samples are filtered on 0.22 µM nitrocellulose filters through Millipore filtration apparatus, and TCA precipitable radioactivity counted in β-counter device after adding the appropriate volume of scintillation liquid. Expected counts may strongly vary depending on different laboratory conditions and reagents. We generally measure 2–5×10^5 cpm of ^{3}H TTP incorporation for 1 µg CAp24 equivalent of VLPs.

3.4. VLP Characterization: Western Blot Analysis

Besides evaluating VLP titers, a critical point is represented by the VLP molecular analysis. In particular, attention should be paid to the efficiency of VLP incorporation of Nef^{mut}-based fusion products. Here, the analyses described below can exhaustively address these requirements from both qualitative and quantitative points of view.

3.4.1. Qualitative Western Blot Analysis

1. Load 200 ng to 1 µg (depending on how efficiently the available antibodies work) CAp24 equivalent of VLPs in four wells of 10–12% SDS-PAGE.

2. Blot gel on filter, cut it in stripes, and reveal VLP products through incubation with anti-HIV-1 Gag, anti-envelope protein, anti-Nef, and anti-heterologous protein moiety Abs.

3.4.2. Semi-Quantitative Western Blot Analysis (see Note 18)

1. Load a unique 10–12% SDS-PAGE gel with serial dilutions of rNef (e.g., five wells containing from 100 to 6.25 ng), and additional five wells with serial dilutions of the VLP preparation, e.g., from 1 µg to 62.5 ng CAp24 equivalent.

2. Incubate the filter with anti-Nef Abs and reveal.

3. Compare the signal intensities from both VLPs and rNef by quantitative densitometry.

3.4.3. Subtilisin A Assay

To ensure that the Nef^{mut}-based fusion product is indeed incorporated, VLPs can be treated with subtilisin A before Western blot assay (see Note 19).

1. Incubate 200 ng to 1 µg of VLPs in a total volume of 30 µl with 1–5 mg/ml of subtilisin A in the appropriate buffer for 15 min at 37°C (see Note 20).

2. Stop the digestion with PMSF to 5 mg/ml.

3. Incubate for 15 min at 95°C.

4. Load on 10–12% SDS-PAGE, blot, and compare the signals from untreated with subtilisin A-treated samples with anti-Nef Abs.

3.5. VLP Characterization: FACS Analysis

In some instances, Abs efficiently recognizing the heterologous moiety of the Nef^{mut} fusion product in Western blot assay are no more available. In case, however, the Abs work well in FACS analysis, a simple and fast FACS-based alternative method for the

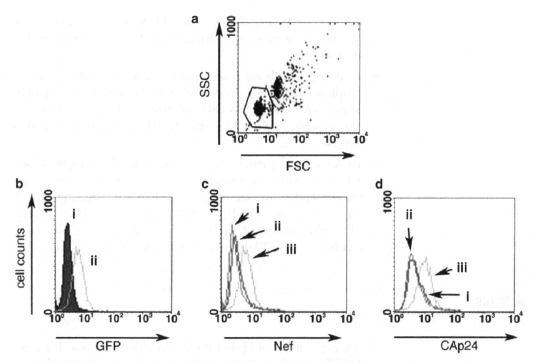

Fig. 2. Cytofluorimetric analysis of aldehyde latex beads-associated Nef^{mut}-GFP VLPs. (**a**) SSC/FSC dot plot of aldehyde latex beads. The gate including single beads is indicated. (**b**) FACS analysis of the VLP contents. 100 μl of supernatant from 293/GPR cells transfected with Nef G3C-GFP vector and induced, were clarified, adjusted to 500 ng of CAp24 contents, incubated with aldehyde latex beads, and analyzed by FACS (*slope ii*). As control, supernatants from noninduced 293/GPR were used (*slope i*). (**c, d**) 500 ng of Nef G3C-GFP VLPs were incubated with aldehyde latex beads, and labeled with either anti Nef mAbs clone 6.2 (**c**) and, in second instance, with PE-conjugated goat anti-mouse IgG, or with KC57 PE-conjugated anti CAp24 mAb (**d**). Histograms refer to FACS analysis of aldehyde latex beads incubated with supernatants of 293/GPR transfected with Nef G3C-GFP vector either noninduced (*slope i*) or induced but in the absence (*slope ii*) or presence (*slope iii*) of the permeabilization step before Ab labeling

detection of VLP products can be performed. It relies on the ability of aldehyde latex beads to specifically bind lipid enveloped micro- and nanoparticles. VLP-coupled beads can be incubated with specific antibodies and analyzed by FACS. This method allows the detection of both membrane associated and intra-particle associated proteins like Gag CAp24 and Nef (Fig. 2).

1. Pretreat the required volume of surfactant-free white aldehyde/ sulfate latex beads (5 μl for up to 5 μg CAp24 equivalent of VLPs) with an excess of FBS at room temperature for 30 min.

2. Wash the beads once with 1× PBS, resuspend them in their original volume, and add 5 μl of beads to either 100 μl of clarified supernatants from VLP producing cells, or 500 ng to 5 μg CAp24 equivalent of concentrated VLP preparations in a final volume of 100 μl of 1× PBS/5% FBS.

3. Incubate at room temperature for 2 h on rotating plate.

4. Wash the beads once with 1 ml of 1× PBS.

5. For direct FACS analysis of GFP fluorescence, resuspend the washed beads in PBS-2% formaldehyde, and perform FACS analysis.

6. To detect membrane associated proteins, incubate the VLP-coupled beads with proper dilution of Abs in 100 μl final volume of 1× PBS/5% FBS, incubate at 4°C for 1 h, wash, resuspend in 1× PBS-2% formaldehyde, and analyze by FACS (see Note 21).

7. To detect intra-particle antigens, treat the VLP-coupled beads with Cytofix/Cytoperm solution according to manufacturer's recommendations (see Note 22). Then, for detecting HIV-1 Gag products, incubate with 1:100 dilution of the PE-conjugated anti-CAp24 KC-57 mAb. For Nef detection, incubate with 1:30 dilution of the anti-Nef mAb clone 6.2.

4. Notes

1. A great number of HIV-1 molecular clones are available from NIH Research and Reference Program. Be sure that the HIV-1 clone of choice contains a full open Nef reading frame, that is not the case of $HTLV_{IIIB}$ derivatives. Alternatively, vectors carrying the HIV-1 wt Nef sequence alone work as well. We use NL4-3-derived Nef isoforms which, differently to SF2-like alleles, lack a 4 amino acid stretch at the N-terminal region.

2. The longest protein we successfully accommodated as C-terminal fusion moiety of Nef^{mut} and incorporated in VLPs was the HCV NS3 protein (i.e., 1,890 nucleotides, 630 amino acids). However, often the VLP incorporation efficiency depends also on intrinsic features of the heterologous polypeptide/protein, such as intracellular localization, hydrophobicity, and interaction with cell membranes. As a general rule, highly hydrophobic amino acid domains which are expected to insert in cell membranes should be deleted from the Nef^{mut}-based fusion protein. In fact, they could interfere with the Nef localization at the inner side of the cell membrane, which is dictated by its N-terminal myristoylation, and enforced by the palmitoylation site created by the G3C mutation.

3. We routinely use combinations of eukaryotic expression vectors regulated by the immediate-early CMV promoter without evident inhibitory interference. However, alternative promoters (e.g., PKG, e-IF2α) have been described working well in 293-derived packaging cells.

4. Depending on the experimental strategy of choice, and also on the available biological material, HIV-1 VLPs can be

recovered from either the inducible 293/GPR packaging cells or 293 T cells. Of course, in the latter case, the co-transfection of HIV-1 packaging construct is required. The experimental procedures are similar in both cases, except that 293 T transfected cells do not require the induction steps. On the other hand, in case the production of MLV-based VLPs is preferred, the Phoenix GP (i.e., 293-based cell line constitutively releasing MLV VLPs, available from ATCC) should be considered as packaging cells.

5. 293/GPR packaging cells are very useful reagent for VLP production. However, particular care should be taken to obtain the best results in terms of cell growth and transfection efficiency. In particular, these cells detach from the solid support very easily, and many floating cells can readily survive. Differently to parental 293 cells, they reach confluence without covering all the available space of the solid support, rather growing in faintly attached clumps. Of note, continuous culturing in the presence of low hygromycin B concentrations (i.e., 0.5–2 µg/ml) guarantees an optimal VLP production at the time of transfection/induction. In any case, do not maintain the cell line for more than 20–30 passages.

6. Do not use UltraClear tubes since the VLP pellet is much more difficult to detach.

7. Several commercial anti-HIV-1 CAp24 ELISA kits are available. In our experience, we found the best efficiency/cost ratio with the kit from Innogenetics.

8. Also in this case, quite reliable kits are available from many companies. However, since our choice has been always directed on an in-house made assay, we have no suggestions regarding the HIV-1 RT kit of preference.

9. We selected HIV-1 strongly positive sera, pooled them, and store at −20°C in 30 µl stocks. This procedure guarantees availability of both reliable and durable reagent over long period of time, considering that 1:1,000/1:3,000 dilutions work quite efficiently in Western blot analysis. Store thawed aliquots at 4°C.

10. There are several good anti-Nef Abs for Western blot both in commerce and from NIH Research and Reference Reagent Program. Among the latter, we recommend also the polyclonal Abs #331 and #2949, and the monoclonal Abs #456 and #1535.

11. It is not easy finding anti-VSV-G Abs working well in Western blot analysis. The Ab from Immunology Consultant Laboratories works very well up to 1:5,000 dilution.

12. Although the foreign antigen incorporated in VLPs is expected to be part of a unique protein fused with Nef, we suggest

to formally confirm the results using also Abs recognizing the heterologous moiety (i.e., anti-GFP Abs in the here described case).

13. The kit from Euroclone is very inexpensive and gives quite good performances.

14. For detecting VLP-associated proteins, the use of monoclonal rather than polyclonal Abs guarantees the best signal/ background ratio. Also, two-steps labeling procedure using fluorochrome-conjugated secondary Abs could increase the sensitivity of the assay.

15. Successful VLP productions can be achieved also by seeding the cells the day of transfection, however using two times more cells, and letting the cells to adhere for at least 3 h.

16. To minimize cell detachment, we routinely dilute the required volumes of both inducers in 2 ml of medium taken from the transfected cultures, and readd the mixtures to the cultures dropwise.

17. This is quite important to maximize VLP recovery. We experienced that also overnight incubation is effective although extending the incubation time beyond 4–6 h does not produce obvious advantages.

18. This assay can be useful to establish the VLP incorporation extents of the Nef^{mut}-based fusion protein and, by consequence, to evaluate possible negative interference of the heterologous moiety. This can be easily assessed considering that Nef^{mut} alone incorporates at 1:2–1:6 molar ratio as compared with CAp24 (26, 33). To evaluate the VLP incorporation efficiency of the Nef^{mut}-based fusion protein, consider also the HIV-1 CAp24 equivalent amounts loaded into the gel, and that HIV-1 particles incorporate approximately 5,000 CAp24 molecules per virion.

19. The choice of appropriate controls for the subtilisin A digestion is critical. In the case of (VSV-G) Nef^{mut}-GFP VLPs, the degradation of the VSV-G ectodomain represents a good control considering that the above quoted anti-VSV-G Abs recognize the intracellular domain of the envelope protein. In this way, the sharp molecular weight reduction of the VSV-G signal in Western blot is good indication that subtilisin A worked properly.

20. Do not use too high concentrations of subtilisin A, or prolong incubation times since undesired, unspecific VLP degradation may occur.

21. The small dimension of the beads implies that these can be detected at quite low values of both SSC/FSC parameters, i.e., similar to those used for detecting platelets.

22. To control that the protein of interest is indeed packed into the viral particles, we suggest to produce control sample using the antibody of interest on nonpermeabilized VLPs-beads complexes.

Acknowledgments

This work was supported by grants from the AIDS project of the Ministry of Health, Rome, Italy. We are indebted to Federica M. Regini for the excellent editorial assistance.

References

1. Grgacic, E. V., and Anderson, D. A. (2006) Virus-like particles: passport to immune recognition. *Methods* **40**, 60–5.

2. Da Silva, D. M., Schiller, J. T., and Kast, W. M. (2003) Heterologous boosting increases immunogenicity of chimeric papillomavirus virus-like particle vaccines. *Vaccine* **21**, 3219–27.

3. Schiller, J. T., and Lowy, D. R. (2001) Papillomavirus-like particle vaccines. *J Natl Cancer Inst Monogr*, 50–4.

4. Newman, M., Suk, F. M., Cajimat, M., Chua, P. K., and Shih, C. (2003) Stability and morphology comparisons of self-assembled virus-like particles from wild-type and mutant human hepatitis B virus capsid proteins. *J Virol* **77**, 12950–60.

5. Zhao, W., Liao, G. Y., Jiang, Y. J., and Jiang, S. D. (2004) Expression and self-assembly of HCV structural proteins into virus-like particles and their immunogenicity. *Chin Med J (Engl)* **117**, 1217–22.

6. Johnson, M. C., Scobie, H. M., Ma, Y. M., and Vogt, V. M. (2002) Nucleic acid-independent retrovirus assembly can be driven by dimerization. *J Virol* **76**, 11177–85.

7. Young, K. R., McBurney, S. P., Karkhanis, L. U., Ross, T. M. (2003) Particle-based vaccines for HIV-1 infection. *Curr Drug Targets Infect Disord* **3**, 151–69.

8. Woo, M. K., An, J. M., Kim, J. D., Park, S. N., and Kim, H. J. (2008) Expression and purification of human papillomavirus 18L1 virus-like particle from saccharomyces cerevisiae. *Arch Pharm Res* **31**, 205–9.

9. Acosta-Rivero, N., Aguilar, J. C., Musacchio, A., Falcón, V., Viña, A., de la Rosa, M. C., and Morales, J. (2001) Characterization of the HCV core virus-like particles produced in the methylotrophic yeast *Pichia pastoris*. *Biochem Biophys Res Commun* **287**, 122–5.

10. Lu, M. W., Liu, W., and Lin, C. S. (2003) Infection competition against grouper nervous necrosis virus by virus-like particles produced in *Escherichia coli*. *J Gen Virol* **84**, 1577–82.

11. Bertolotti-Ciarlet, A., Ciarlet, M., Crawford, S. E., Conner, M. E., and Estes, M. K. (2003) Immunogenicity and protective efficacy of rotavirus 2/6-virus-like particles produced by a dual baculovirus expression vector and administered intramuscularly, intranasally, or orally to mice. *Vaccine* **21**, 3885–90.

12. Buonaguro, L., Tornesello, M. L., Tagliamonte, M., Gallo, R. C., Wang, L. X., Kamin-Lewis, R., Abdelwahab, S., Lewis, G. K., and Buonaguro, F. M. (2006) Baculovirus-derived human immunodeficiency virus type 1 virus-like particles activate dendritic cells and induce ex vivo T-cell responses. *J Virol* **80**, 9134–43.

13. Ramqvist, T., Andreasson, K., and Dalianis, T. (2007) Vaccination, immune and gene therapy based on virus-like particles against viral infections and cancer. *Expert Opin Biol Ther* **7**, 997–1007.

14. McAleer, W. J., Buynak, E. B., Maigetter, R. Z., Wampler, D. E., Miller, D. E., and Hilleman, M. R. (1984) Human hepatitis B vaccine from recombinant yeast. *Nature* **307**, 178–80.

15. Garland, S. M., Hernandez-Avila, M., Wheeler, C. M., Perez, G., Harper, D. M., Leodolter, S., Tang, G. W., Ferris, D. G., Steben, M., and Bryan, J. (2007) Females united to unilaterally reduce endo/ectocervical disease (FUTURE) I investigators: quadrivalent vaccine against human papillomavirus to prevent anogenital diseases. *N Engl J Med* **356**, 1928–43.

16. Malboeuf, C. M., Simon, D. A., Lee, Y. E., Lankes, H. A., Dewhurst, S., Frelinger, J. G., Rose, R.C. (2007) Human papillomavirus-like particles mediate functional delivery of plasmid DNA to antigen presenting cells in vivo. *Vaccine* **25**, 3270–6.

17. Wu, M., Sherwin, T., Brown, W.L., Stockley, P.G. (2005) Delivery of antisense oligonucleotides to leukemia cells by RNA bacteriophage capsids. *Nanomedicine* **1**, 67–76.

18. Wang, Q., Lin, T., Tang, L., Johnson, J.E., and Finn, M.G. (2002) Icosahedral virus particles as addressable nanoscale building blocks. *Angew Chem Int Ed Engl* **41**, 459–62.

19. Gleba, Y., Klimyuk, V., and Marillonnet, S. (2007) Viral vectors for the expression of proteins in plants. *Curr Opin Biotechnol* **18**, 134–41.

20. Mao, C., Solis, D.J., Reiss, B.D., Kottmann, S.T., Sweeney, R.Y., Hayhurst, A., Georgiou, G., Iverson, B., and Belcher, A.M. (2004) Virus-based toolkit for the directed synthesis of magnetic and semiconducting nanowires. *Science* **303**, 213–7.

21. Montefiori, D.C., Safrit, J.T., Lydy, S.L., Barry, A.P., Bilska, M., Vo, H.T., Klein, M., Tartaglia, J., Robinson, H.L., and Rovinski, B. (2001) Induction of neutralizing antibodies and gag-specific cellular immune responses to an R5 primary isolate of human immunodeficiency virus type 1 in rhesus macaques. *J Virol* **75**, 5879–90.

22. Deml, L., Kratochwil, G., Osterrieder, N., Knuchel, R., Wolf, H., and Wagner, R. (1997) Increased incorporation of chimeric human immunodeficiency virus type 1 gp120 proteins into Pr55gag virus-like particles by an Epstein-Barr virus gp220/350-derived transmembrane domain. *Virology* **235**, 10–25.

23. Kattenbeck, B., Von Poblotzki A., Rohrhofer, A., Wolf, H., and Modrow, S. (1997) Inhibition of human immunodeficiency virus type 1 particle formation by alterations of defined amino acids within the C terminus of the capsid protein. *J Gen Virol* **78**, 2489–96.

24. Muller, B., Daecke, J., Fackler, O. T., Dittmar, M. T., Zentgraf, H., and Krausslich, H. G. (2004) Construction and characterization of a fluorescently labeled infectious human immunodeficiency virus type 1 derivative. *J Virol* **78**, 10803–13.

25. Wu, X., Liu, H., Xiao, H., Kim, J., Seshaiah, P., Natsoulis, G., Boeke, J. D., Hahn, B. H., and Kappes, J. C. (1995) Targeting foreign proteins to human immunodeficiency virus particles via fusion with Vpr and Vpx. *J Virol* **69**, 3389–98.

26. Peretti, S., Schiavoni, I., Pugliese, K., and Federico, M. (2005) Cell death induced by the herpes simplex virus-1 thymidine kinase delivered by human immunodeficiency virus-1-based virus-like particles. *Mol Ther* **12**, 1185–96.

27. Yao, X. J., Kobinger, G., Dandache, S., Rougeau, N., and Cohen, E. (1999) HIV-1 Vpr-chloramphenicol acetyltransferase fusion proteins: sequence requirement for virion incorporation and analysis of antiviral effect. *Gene Ther* **6**, 1590–9.

28. McDonald, D., Vodicka, M. A., Lucero, G., Svitkina, T. M., Borisy, G. G., Emerman, M., Hope, T. J. (2002) Visualization of the intracellular behavior of HIV in living cells. *J Cell Biol* **159**, 441–52.

29. Cavrois, M., Noronha, C., and Greene, W. C. (2002) A sensitive and specific enzyme-based assay detecting HIV-1 virion fusion in primary T lymphocytes. *Nat Biotechnol* **20**, 1151–4.

30. Muratori, C., D'Aloja, P., Superti, F., Tinari, A., Sol-Foulon, N., Sparacio, S., Bosch, V., Schwartz, O., and Federico, M. (2006) Generation and characterization of a stable cell population releasing fluorescent HIV-1-based Virus Like Particles in an inducible way. *BMC Biotechnol* **6**, 52.

31. Krautkramer, E., Giese, S. I., Gasteier, J. E., Muranyi, W., and Fackler, O. T. (2004) Human immunodeficiency virus type 1 Nef activates p21-activated kinase via recruitment into lipid rafts. *J Virol* **78**, 4085–97.

32. Mañes, S., del Real, G., Lacalle, R. A., Lucas, P., Gómez-Moutón, C., Sánchez-Palomino, S., Delgado, R., Alcamí, J., Mira, E., and Martínez, A.C. (2000) Membrane raft microdomains mediate lateral assemblies required for HIV-1 infection. *EMBO Rep* **1**, 190–6.

33. Peretti, S., Schiavoni, I., Pugliese, K., and Federico, M. (2006) Selective elimination of HIV-1-infected cells by Env-directed, HIV-1-based virus-like particles. *Virology* **345**, 115–26.

34. Green, L.A., Lu, Y., and He, J.J. (2009). Inhibition of HIV-1 infection and replication by enhancing viral incorporation of innate anti-HIV-1 protein A3G. *J Biol Chem* **284**, 13363–72.

35. Sparacio, S., Pfeiffer, T., Schaal, H., and Bosch, V. (2001) Generation of a flexible cell line with regulatable, high-level expression of HIV Gag/Pol particles capable of packaging HIV-derived vectors. *Mol Ther* **3**, 602–12.

36. Geyer, M., Fackler, O. T., and Peterlin, B. M. (2001) Structure-function relationships in HIV-1 Nef. *EMBO Rep* **2**, 580–5.

37. D'Aloja, P., Santarcangelo, A. C., Arold, S., Baur, A. and Federico, M. (2001) Genetic and functional analysis of the human immunodeficiency virus (HIV) type 1-inhibiting F12-HIVnef allele. *J Gen Virol* **82**, 2735–45.

Part III

New Lentiviral Vector Applications

Chapter 8

Lentiviral Vector-Mediated Transgenesis in Human Embryonic Stem Cells

Zhong-Wei Du and Su-Chun Zhang

Abstract

Human Embryonic stem cells (hESCs) offer an invaluable tool for revealing human biology and a potential source of functional cells/tissues for regenerative medicine. The utility of hESCs will likely be significantly enhanced and broadened by our ability to build versatile genetically modified hESC lines. Here, we describe an efficient lentiviral vector mediated method to establish stable transgenic hESCs.

Key words: Human embryonic stem cell, Lentivirus, Transgenic, Promoter, Bicistronic

1. Introduction

Human embryonic stem cells (hESCs) are derived from the blastocyst embryo (1, 2), which can be propagated unlimitedly in vitro and have the potential to produce any cell types of the human body. Thus, hESCs offer an invaluable tool for revealing human biology and a potential source of functional cells/tissues for regenerative medicine.

Like their mouse counterparts, which have revolutionized biomedical research through transgenesis in the past few decades (3, 4), the utility of hESCs will likely be significantly enhanced and broadened by our ability to build versatile genetically modified hESC lines (5, 6). The transgenic hESCs may be undertaken to optimize the directed differentiation toward a specific cell lineage (7), facilitate the purification of cell populations by introduction of a fluorescent reporter gene (8), or correct genetic defects by ectopic expression or silencing of specific genes (9).

Building stable transgenic hESC lines remains a challenging process. Transgenes can be introduced into hESCs by plasmid

Maurizio Federico (ed.), *Lentivirus Gene Engineering Protocols,* Methods in Molecular Biology, vol. 614,
DOI 10.1007/978-1-60761-533-0_8, © Humana Press, a part of Springer Science+Business Media, LLC 2010

transfection using electroporation and chemical based transfection reagents, which however leads to low transfection efficiency with 1 stably transfected cell per 10^5 cells (10, 11). To date, lentiviral vector-mediated transfection is a preferred strategy for the following advantages over other methods: (1) It offers a high transfection efficiency, up to 70% being achieved with the high titer of self-inactivating lentivirus (12); (2) It is less prone to gene silencing as the transgenes integrate permanently into gene expression hotspots (13), and (3) It sustains long term transgene expression during cell passaging and ESC differentiation (14).

When applying lentiviral vector-mediated transgenesis in hESCs, the most critical issue to be considered is the selection of a promoter. We have tested four ubiquitous constitutive promoters in the lentiviral vector to drive green fluorescent protein (GFP) gene in hESCs: cytomegalovirus (CMV), phosphoglycerate kinase (PGK), cytomegalovirus immediate-early enhancer/chicken β-actin hybrid (CAG) and elongation factor-1α (EF1α). We found that the expression of GFP is suppressed in a promoter-dependent manner (15). The CMV promoter yields the lowest percentage (1.1%) of transfection and the lowest level of GFP expression: thus, it is not suitable for transgenesis in human ESCs. The PGK promoter yields a high percentage of transfection, but a medium level of GFP expression. The EF1α promoter is preferred in hESCs as it produces the highest level of GFP expression with reasonable transfection efficiency. However, EF1α promoter-driven transgene (GFP) expression tends to be turned off following long-term cell differentiation, especially in neurons (16). The CAG promoter seems a better choice for sustained transgene expression during cell differentiation (17). The transcriptional activities of the above ubiquitous promoters were addressed in hESCs, which might be different in other human cell types. Therefore, one should test the promoter activity in the target cell first. Another critical issue is the selection method for obtaining a pure population of transfected cells. For ubiquitous constitutive promoters, one potential approach is to use bicistronic lentivectors, containing both a gene of interest and a selection marker under the control of the same promoter. For cell-type specific promoters, the selection marker needs to be driven by a separate constitutive promoter for selection at the hESC stage. Here, we present a detailed protocol for establishing stable transgenic hESC lines by bicistronic lentiviral vector-mediated transfection.

2. Materials

2.1. Lentivirus Preparation by Calcium–Phosphate Transfection

1. Lentivirus plasmid DNA (see Note 1).

2. Packaging plasmid psPAX2 (Addgene, Cambridge, MA, plasmid 12260).

3. Envelope plasmid pMD2.G (Addgene, Cambridge, MA, plasmid 12259).

4. 2× BES buffered saline (Sigma-Aldrich, St. Louis, MO, Cat. 14280).

5. 1 M $CaCl_2$.

6. 0.1% gelatin: 1 g gelatin in 1,000 mL dH_2O, sterilized and store at RT.

7. Packaging cell line 293T (ATCC, Manassas, VA, CRL-11268).

8. Culture medium: For 293T cells use DMEM (Dulbecco's Modified Eagle Medium, Invitrogen, Carlsbad, CA, Cat. 11965) containing 10% fetal bovine serum (FBS).

9. Collection medium – hESC medium (see Note 2): DMEM/F12 (Invitrogen, Cat. 11320-033), 20% (v/v) knockout serum replacement (Invitrogen, Cat. 10828-028), 10 mM nonessential amino acids (Invitrogen, Cat.11140-050) 2 mM l-glutamine (Invitrogen, Cat. 25030-081), 50 mM β-mercaptoethanol.

10. 0.45 μm Steriflip® (Millipore, Billerica, MA, Cat. SE1M-003M00).

2.2. hESC Culture on Matrigel

1. Matrigel: BD Matrigel™ hESC-qualified Matrix (BD Bioscience, San Jose, CA, Cat. 354277); aliquot Matrigel at 2 mg/tube into sterile Eppendorf tube at 4°C, and store at –20°C. Each tube can be used for coating one 6-well plate.

2. hESC lines: H9 or H1 cell lines (National stem cell bank, Madison, WI, http://www.wicell.org/, NIH registered code WA09, WA01).

3. Dispase: dissolve dispase (Invitrogen, Cat. 17105-041) at 1 mg/mL in DMEM/F12 (Invitrogen, Cat. 11320-033) medium, warm at 37°C for 10 min to completely dissolve and filter sterile with 0.22 μm Steriflip® (Millipore, Cat. SCGP00525). Store at 4°C and use within 2 weeks.

4. Conditioned medium (CM): 15 mL hESC medium is conditioned on irradiated MEFs (mouse embryonic fibroblasts, 3×10^6 cells in T75 flask) and collected every 24 h. 4 ng/mL bFGF is added immediately before using on hESCs. The MEFs can last for 2 weeks for CM.

2.3. Lentiviral Vector Transfection

1. Y-27632 (Calbiochem, San Diego, CA, Cat. 688000): dissolve 1 mg Y-27632 in 296 μL sterile water to make 1,000× stock, aliquot 50 μL/tube and store at –20°C.

2. TrypLE select (Invitrogen, Cat. 12563-029).

3. Blasticidin S (Invivogen, San Diego, CA, Cat. ant-bl-1, 10 mg/mL).

2.4. Amplification and Verification of Transgenic Clones

1. MEF feeder plate: prepare irradiated MEFs at 0.75×10^5 cells/mL, plate on gelatin-coated plate at 0.5 mL/well of 24-well plate and 2.5 mL/well of 6-well plate, and allow the MEFs to attach overnight in the incubator. MEF feeder plate can be used within 1 week and needs to be washed once with hESC medium before applying hESCs.

3. Methods

3.1. Lentivirus Preparation by Calcium–Phosphate Transfection

1. Coat 10-cm cell culture dish with 10 mL 0.1% gelatin, place at 4°C for 1 h (see Note 3).

2. Aspirate off the gelatin, plate 2.5×10^6 of 293T cells per 10-cm dish in 10 mL culture medium and incubate at 37°C with 5% CO_2.

3. 24 h later, prepare calcium–phosphate precipitate (1 mL/10 cm dish): mix 20 µg lentivirus plasmid, 15 µg packaging plasmid and 6 µg envelope plasmid in a 1.5 mL sterile Eppendorf tube (see Note 4). Add 125 µL 1 M $CaCl_2$ and bring the volume to 0.5 mL with sterile water. Add 0.5 mL $2 \times BES$ buffered saline and shake briefly. Keep in RT for 20 min. Add dropwise to the cells and mix gently with medium. Incubate at 37°C with 5% CO_2 (see Note 5).

4. Change medium (6–8 h later) by removing the medium with precipitate and add 6 mL/dish of collection medium (see Note 6).

5. Collect medium 48 h later to a 15 mL conical tube and spin at $2,000 \times g/5$ min/RT. Filter the supernatant through a 0.45 µm Steriflip®. At this point, lentivirus can be used directly for hESC transfection, or concentrated for long-term storage at −70°C (see Note 7).

3.2. hES Cell Culture on Matrigel (see Note 8)

1. Thaw Matrigel by hand, dilute with 6 mL cold DMEM/F12 in a 15 mL tube right after thawing and mix well. Add 1 mL/well of a 6-well plate and place the plate at 4°C overnight (see Note 9).

2. Aspirate medium from ESC culture, add 2 mL dispase/well of a 6-well plate. Incubate at 37°C for 3–5 min until edges of colonies begin to curl up. Remove the dispase and add 3 mL hESC medium/well to the plate (see Note 10).

3. Collect cells by washing and scraping with the hESC medium, then transfer to a 15 mL tube. Spin down at $200 \times g/3$ min/RT and resuspend in 3 mL conditioned medium (CM).

4. Aspirate Matrigel coating medium from a 6 well plate and add 2 mL CM to each well. Pipette up and down to mix cells first,

then add 0.5 mL cell suspension to one well. Repeat this step to add 0.5 mL cell suspension to each well (see Note 11).

5. Place gently into incubator at 37°C with 5% CO_2. Make sure the cells are evenly distributed across the plate by gently shaking the plate side to side, then front to back (see Note 12).

6. Change medium with 2.5 mL CM/well daily and passage the hESCs every 5–6 days.

3.3. Lentiviral Vector Transfection of hESC

1. Coat a 6-well plate with Matrigel as described in step 1 of Subheading 3.2.

2. Aspirate medium from two wells of a 6-well plate of ESC culture; add 2 mL CM/well with 10 μM Y-27632 and incubate at 37°C for 1 h (see Note 13).

3. Aspirate off the CM and Y-27632, add 2 mL TrypLE select/well.

4. Incubate at 37°C for 5 min, then remove TrypLE select and add 3 mL hESC medium/well.

5. Collect the hESCs by washing with the hESC medium, then transfer to a 15 mL tube. Spin down at $480 \times g$/3 min/RT and then aspirate the supernatant.

6. Resuspend in 1 mL lentivirus containing medium from step 5 of Subheading 3.1 (or 1 mL CM plus 10 μL of concentrated lentiviral vector preparation). Incubate at 37°C for 1 h (see Note 14).

7. Bring the volume with CM to 15 mL and add 10 μM Y-27632. Mix and resuspend the hESCs.

8. Aspirate matrigel coating medium from a 6 well plate and plate 2.5 mL cell suspension into each well.

9. Place gently into incubator. Make sure the cells are evenly distributed across the plate by gently shaking the plate side to side, then front to back.

10. Change medium with 2.5 mL CM/well daily for 5 days. Then, apply selection drug Blasticidin S at 5 μg/mL in the CM, change medium daily and select for 2 weeks (see Note 15).

3.4. Amplification and Verification of Transgenic Clones

1. Treat the hESC colonies with dispase as in steps 2 and 3 of Subheading 3.2.

2. Use a 100 μL pipette tip to transfer each colony into 100 μL hESC medium in a 0.5-mL Eppendorf tube. Triturate up and down 3–5 times to break the colony into small pieces, and then transfer to 24-well plate with MEF feeder containing 0.5 mL hESC medium/well.

3. Culture for 1 week, then passage the cells from the original colony into duplicate two wells of 6 well plate. One well of

cells can be frozen for storage in liquid nitrogen, the other well can be used to determine the transgene expression level by PCR, Western blot, or immunostaining analysis.

4. Notes

1. We use the lentiviral vector system developed by Prof. Didier Trono's lab. The lentivirus DNA containing CAG, EF1α and PGK promoters can be obtained from Addgene (plasmid 11643, 11644, 11642). We replace the IRES-tTRKRAB with IRES-Bsr (Blasticidin S resistant gene) and the GFP gene with the gene of interest.

2. The 293T cell culture medium contains fetal bovine serum which interferes with the undifferentiated state of hESCs. We thus use hESC medium to collect the lentivirus.

3. The 293T cells tend to detach from the cell culture dish during $CaPO_4$ transfection and subsequent lentiviral vector collection. Hence, we use gelatin to improve cell attachment.

4. Plasmids can be sterilized by ethanol precipitation and dissolved in sterile water.

5. Shake the cell culture dish gently to distribute the calcium–phosphate mixture evenly in the whole dish, but be careful as over shaking can detach the cell monolayer.

6. Watch the cells carefully between 6 and 8 h, and change the medium if the cells begin to shrink. Change medium gently as the cells are easy to detach at this stage.

7. You can determine the titer of lentivirus by transfecting 293T cells or commercial kits. We mostly obtained about 10^7 TU/mL with this protocol, which is about 5–10 MOI to hESCs. For long-term storage, transfer 30 mL of virus to 33 mL Beckman conical tubes and spin at $47,000 \times g/2$ h/4°C in Beckman SW28 swinging bucket rotor. After spin, discard the supernatant and resuspend the lentiviral vector preparation in 300 μL of PBS/1% BSA. Aliquot and store at −70°C.

8. MEF feeder cells absorb lentivirus more efficiently than hESCs, therefore we passage hESCs on Matrigel once to get rid of MEFs and perform lentivirus transfection experiments under feeder-free condition.

9. Always prepare the Matrigel at cold temperatures and quickly, as the Matrigel begins to solidify at RT. Once solidified the Matrigel cannot be used again.

10. Watch the hESCs carefully, and remove the dispase right away when edges of the ESC colonies begin to curl up. It is difficult to break the colonies into small pieces if overdigested with dispase.

11. To plate hESCs evenly to each well of the 6-well plate, pipette up and down to mix the cells before adding the cells to each well.

12. Distribute the cells gently and evenly. When using quick motions, you will most likely wind up with your cells in the middle of the plate, or splashed onto the cover. When moving your plate in the incubator to disperse the cells, be sure to move only in one direction at a time.

13. Y-27632 is a chemical of the ROCK inhibitor family, which is applied to improve the survival rate when hESCs are dissociated into single cells or small pieces (18).

14. Incubating with high concentration of lentivirus for 1 h can improve the transfection efficiency, but extending the incubation may induce differentiation of hESCs.

15. After 2 weeks' selection, the transfected colonies grew to 2 μm diameter in size, which can easily be seen by the naked eye.

Acknowledgments

We thank Prof. Didier Trono for supplying lentiviral vectors and Prof. Keiya Ozawa for supplying the plasmid pEB2 containing IRES-Bsr.

References

1. Thomson, J. A., Itskovitz-Eldor, J., Shapiro, S. S., Waknitz, M. A., Swiergiel, J. J., Marshall, V. S., and Jones, J. M. (1998) Embryonic stem cell lines derived from human blastocysts. *Science* **282**, 1145–7.

2. Reubinoff, B. E., Pera, M. F., Fong, C. Y., Trounson, A., and Bongso, A. (2000) Embryonic stem cell lines from human blastocysts: somatic differentiation in vitro. *Nat Biotechnol* **18**, 399–404.

3. Downing, G. J., and Battey, J. F., Jr. (2004) Technical assessment of the first 20 years of research using mouse embryonic stem cell lines. *Stem Cells* **22**, 1168–80.

4. Rossant, J., Bernelot-Moens, C., and Nagy, A. (1993) Genome manipulation in embryonic stem cells. *Philos Trans R Soc Lond B Biol Sci* **339**, 207–15.

5. Xia, X., and Zhang, S. C. (2007) Genetic modification of human embryonic stem cells. In: Biotechnology and genetic engineering reviews, Vol 24, Harding, S.E., eds. Nottingham University Press, Nottingham, 297–310.

6. Zeng, X., and Rao, M. S. (2008) Controlled genetic modification of stem cells for developing drug discovery tools and novel therapeutic applications. *Curr Opin Mol Ther* **10**, 207–13.

7. Seguin, C. A., Draper, J. S., Nagy, A., and Rossant, J. (2008) Establishment of endoderm progenitors by SOX transcription factor expression in human embryonic stem cells. *Cell Stem Cell* **3**, 182–95.

8. Singh, R. N., Nakano, T., Xuing, L., Kang, J., Nedergaard, M., and Goldman, S. A. (2005) Enhancer-specified GFP-based FACS purification of human spinal motor neurons from embryonic stem cells. *Exp Neurol* **196**, 224–34.

9. Chang, J. C., Ye, L., and Kan, Y. W. (2006) Correction of the sickle cell mutation in embryonic stem cells. *Proc Natl Acad Sci U S A* **103**, 1036–40.

10. Eiges, R., Schuldiner, M., Drukker, M., Yanuka, O., Itskovitz-Eldor, J., and Benvenisty, N. (2001) Establishment of human embryonic stem cell-transfected clones carrying a marker for undifferentiated cells. *Curr Biol* **11**, 514–8.

11. Zwaka, T. P., and Thomson, J. A. (2003) Homologous recombination in human embryonic stem cells. *Nat Biotechnol* **21**, 319–21.

12. Xiong, C., Tang, D. Q., Xie, C. Q., Zhang, L., Xu, K. F., Thompson, W. E., Chou, W., Gibbons, G. H., Chang, L. J., Yang, L. J., and Chen, Y. E. (2005) Genetic engineering of human embryonic stem cells with lentiviral vectors. *Stem Cells Dev* **14**, 367–77.

13. Schroder, A. R., Shinn, P., Chen, H., Berry, C., Ecker, J. R., and Bushman, F. (2002) HIV-1 integration in the human genome favors active genes and local hotspots. *Cell* **110**, 521–9.

14. Pfeifer, A., Ikawa, M., Dayn, Y., and Verma, I. M. (2002) Transgenesis by lentiviral vectors: lack of gene silencing in mammalian embryonic stem cells and preimplantation embryos. *Proc Natl Acad Sci U S A* **99**, 2140–5.

15. Xia, X., Zhang, Y., Zieth, C. R., and Zhang, S. C. (2007) Transgenes delivered by lentiviral vector are suppressed in human embryonic stem cells in a promoter-dependent manner. *Stem Cells Dev* **16**, 167–76.

16. Guillaume, D. J., Johnson, M. A., Li, X. J., and Zhang, S. C. (2006) Human embryonic stem cell-derived neural precursors develop into neurons and integrate into the host brain. *J Neurosci Res* **84**, 1165–76.

17. Hong, S., Hwang, D. Y., Yoon, S., Isacson, O., Ramezani, A., Hawley, R. G., and Kim, K. S. (2007) Functional analysis of various promoters in lentiviral vectors at different stages of in vitro differentiation of mouse embryonic stem cells. *Mol Ther* **15**, 1630–9.

18. Watanabe, K., Ueno, M., Kamiya, D., Nishiyama, A., Matsumura, M., Wataya, T., Takahashi, J. B., Nishikawa, S., Nishikawa, S., Muguruma, K., and Sasai, Y. (2007) A ROCK inhibitor permits survival of dissociated human embryonic stem cells. *Nat Biotechnol* **25**, 681–6.

Chapter 9

Lentiviral Vector Transduction of Fetal Mesenchymal Stem Cells

Mark S. K. Chong and Jerry Chan

Abstract

Human fetal mesenchymal stem cells (hfMSC) demonstrate extensive expansion and differentiation capacities and are hence being studied for use in stem cell therapeutics, including gene delivery. With advanced prenatal diagnosis, fetal gene therapy represents an additional avenue for the treatment of inherited deficiencies. We have recently demonstrated harvesting of first-trimester fMSC from fetal blood for ex vivo genetic engineering to introduce genes of interest, and finally intra-uterine transplantation (IUT) of these cells to the fetus. Here we discuss methods in the harvesting of hfMSC, lentiviral transduction to introduce genes of interest, and in vitro methods to characterise transgene expression.

Key words: Human fetal mesenchymal stem cells, Ex vivo transduction, Fluorescence microscopy, Flow cytometry, Osteogenic induction, Adipogenic induction, Chondrogenic induction

1. Introduction

1.1. MSC and hfMSC

Mesenchymal stem cells (MSC), also known as marrow stromal cells or colony forming unit-fibroblasts (CFU-F), were first isolated from adult bone marrow by Friedenstein et al. in the 1970s (1). Other sources of MSC have since emerged, including those from adipose tissues, umbilical cord, umbilical-cord blood, and even dental pulp (2–6). Human fetal MSC (hfMSC) have similarly been isolated from blood, liver and bone marrow in the first trimester fetus, and demonstrated several advantages over adult and perinatal sources (7, 8). Like their adult counterparts, hfMSC are capable of self-renewal and differentiation toward multiple lineages, including bone, fat, and cartilage (7). However, hfMSC proliferate faster, undergo more population doublings, have longer telomeres and express greater levels of telomerase (9),

Maurizio Federico (ed.), *Lentivirus Gene Engineering Protocols,* Methods in Molecular Biology, vol. 614,
DOI 10.1007/978-1-60761-533-0_9, © Humana Press, a part of Springer Science+Business Media, LLC 2010

and demonstrate greater plasticity, differentiating into both skeletal myocytes and oligodendrocytes (8, 10). hfMSC are HLA class II negative, express low levels of HLA class I, and possess immuno-modulatory activity (11–13). The infusion of fully allogeneic hfMSC into a human fetus with skeletal dysplasia led to significant chimerism and possible clinical benefit, without evidence of immune rejection (14). Thus, hfMSC represent a potential cell source for various clinical applications, ranging from cellular replacement therapy and tissue engineering, to cellular-based gene therapy (8, 13, 15, 16).

1.2. MSC as Targets of Gene Transfer

MSC have been found to be capable of homing and migrating to areas of tissue damage (16–21), suggesting their utility for both cellular replacement and well as carriers for transgenes. Genetically modified MSC have now been investigated for (a) Treatment of degenerative diseases, including the delivery of therapeutic factors to diseased tissue; (b) Therapy of single-factor inherited deficiencies, including hemophilia; (c) Delivery of anti-metastatic genes to tumors; and (d) Cellular sources for tissue engineering applications with improved proliferative or differentiation potential (22, 23).

With the convergence of molecular techniques and prenatal diagnosis technologies, early prenatal diagnosis of various inherited genetic disorders can be achieved early in the first trimester, leading to the possibility of effecting gene therapy during a critical period of fetal immune naïveté (24–27). We recently demonstrated the harvest of first trimester hfMSC through high resolution ultrasound directed or embryo-fetoscopic directed sampling of extra-fetal vessels, raising the possibility of an ex vivo gene therapy for autologous intra-uterine transplant (8, 15). An autologous approach would obviate immune responses associated with the use of allogeneic cell sources, and induce central tolerance to the foreign gene product when performed within a critical window of around 14 weeks' gestation in humans (24, 28, 29).

Retroviral vectors are favored in such applications for integration into the host genome and consequently, capacity for stable expression. However, commonly used onco-retroviral vectors exhibit poor gene transfer efficiency, resulting in poor clinical outcomes (30). Similarly, we have found superior transduction rates and duration of expression with the use of lentiviral constructs when compared with a Moloney Leukaemia Viral (MLV) construct (31). Lentiviral transduction of hfMSC did not affect its proliferation or multi-lineage differentiation capacity, with stable expression over 50 population doublings, highlighting the feasibility of hfMSC as a target for ex vivo gene manipulation. hfMSC have also been investigated in a fetal-to-fetal xenogeneic transplantation paradigm in fetal mice models of muscular dystrophy and skeletal dysplasia. Intrauterine transplantation of hfMSC in these mice led to widespread engraftment and site specific differentiation in multiple tissue types with preferred homing to injured tissues, while maintaining transgene expression for up to 19 weeks duration (16, 19). We are currently studying the transduction of hfMSC for the treatment of single gene disorders such as hemophilia B.

In this chapter, we detail the harvesting of hfMSC from fetal bone, liver and blood. We further describe the lentiviral transduction of hfMSC, and assays for the detection of the expression of the gene of interest. These protocols may have applications in the genetic engineering of hfMSC for a variety of applications such as ex vivo gene therapies involving the delivery of transgenes in the setting of inherited or degenerative diseases.

2. Materials

All reagents were obtained from Sigma-Aldrich company, St. Louis, MO, unless otherwise stated.

2.1. Cell Harvesting and Culture

1. Phosphate-buffered saline (PBS).
2. 70-μm nylon mesh cell strainer (BD Biosciences, San Diego, CA).
3. Ficoll-Paque Plus (GE Healthcare, Piscataway, NJ).
4. D10: High glucose Dulbecco's Modified Eagle's Medium (DMEM-HG) supplemented with Glutamax, 50 U/ml penicillin, 50 μg/ml streptomycin and 10% fetal bovine serum (all from Gibco, Carlsbad, CA).
5. Gentamycin Sulfate (Invitrogen).

2.2. Lentivector Preparation and Transduction

1. Calcium phosphate transfection kit (Invitrogen, Carlsbad, CA).
2. Vector plasmids (see Note 1).
3. 293T cells (ATCC, Manassas, VA).
4. Polybrene (Hexadimethrine bromide): 8 mg/ml in H_2O (Stock solution).

2.3. Characterization of Gene of Interest

2.3.1. Fluorescence-Activated Cell Sorting

1. Fixative: 0.01% formaldehyde in PBS.
2. Permeabilising medium: 1% Triton-X100 in PBS.
3. FACs Buffer: Permeabilising medium supplemented with 1% BSA (fluorescence-activated cell sorting (FACs) buffer).
4. Primary antibody.

2.3.2 Immunostaining

1. Fixative: 1:1 methanol/acetone solution.
2. Blocking solution:

   (a) 5% w/v BSA.

   (b) 2% w/v Goat Serum.

   (c) 0.3% v/v Triton-X100.

   (d) PBS.

3. Primary antibody solution:

   (a) 10% Blocking solution in PBS.

   (b) Primary antibody (see Note 2).

4. Secondary antibody solution:

   (a) 1% Bovine Serum Albumin in PBS.

   (b) Fluorescence conjugated secondary antibody (see Note 3).

2.4. Induction of Multilineage Differentiation

Supplements obtained from Sigma unless otherwise stated.

1. 10% formalin.

2. Bone differentiation medium (see Note 4):

   (a) D10.

   (b) 10 mM β-glycerophosphate.

   (c) 0.01 μM Dexamethasone.

   (d) 0.2 mM Ascorbic acid.

3. Fat supportive medium

   (a) D10.

   (b) M Insulin.

4. Fat supportive medium:

   (a) 1 μM dexamethasone.

   (b) 0.2 mM indomethacin.

   (c) 0.5 mM isobutylmethylxanthine (IBMX).

5. Cartilage differentiation medium:

   (a) High glucose Dulbecco's Modified Eagle's Medium (DMEM-HG).supplemented with Glutamax, 50 U/ml penicillin, 50 μg/ml streptomycin (all from Gibco).

   (b) 0.1 mM Dexamethasone.

   (c) 0.17 mM Ascorbic acid 2-phosphate.

   (d) 1 mM Sodium Pyruvate.

   (e) 0.35 mM L-Proline.

   (f) Insulin–Transferrin–Selenite supplements.

       – 0.25 mg/ml insulin.

       – 0.25 mg/ml transferring.

       – 0.25 μg/ml selenite.

   (g) 0.01 g/ml Bovine Serum Albumin.

(h) 0.006 μg/ml Linoleic acid.

(i) 1 μg/ml TGF-β3.

2.5. Characterization of Multilineage Differentiation

1. 10% Formalin.

2. Deionised water (dH$_2$O).

3. Histoclear.

2.5.1. Characterization of Bone Tissue

1. 2% Silver nitrate in dH$_2$O.

2. 1% Pyrogallol acid in dH$_2$O.

3. 15% sodium thiosulfate in dH$_2$O.

2.5.2. Characterization of Adipose Tissue

1. 1% Oil Red O stock solution

   a. 1 g Oil Red O.

   b. 100 ml isopropanol (or Triethyl phosphate, for working solution, dilute in same).

2. Oil Red O working solution

   (a) Parts of 1% Oil Red O stock solution.

   (b) Two parts dH$_2$O (6 ml Oil Red O + 4 ml dH$_2$O).

2.5.3. Characterization of Cartilage Tissue

1. 0.1% Safranin O Solution in dH$_2$O.

2. Ethanol.

3. Methods

3.1. Isolation of Cell Suspensions from Fetal Tissue

Fetal tissue should be collected in compliance with local ethical guidelines. Tissue samples should be processed within 4 h of collection, and are to be prepared within a suitable biological safety cabinet.

3.1.1. Bone Marrow

1. Dissect out long bones (femur and humerus).

2. Carefully remove muscle tissue to prevent myoblast contamination.

3. Dissect out shaft. Carefully flush out bone marrow with 20 ml of DMEM using a 23G needle and pass through 70 μm cell strainer.

3.1.2. Liver

1. Remove Gilsson's capsule to reduce mesothelial contamination.

2. Mince liver tissue with scalpel in DMEM.

3. Mechanically triturate sequentially through 5, 2, and 1 ml pipettes, then through 18G, 20G needles.

4. Pass triturate through 70 μm cell strainer.

3.1.3. Blood

1. Aspirate blood into 5 ml syringe containing 100 µl of heparin solution via cardiac puncture.

2. Deplete red blood cells using Ficoll-Paque as per manufacturer's instructions.

3. Carefully collect interphase and resuspend in DMEM.

3.2. Selection of Human Fetal Mesenchymal Stem Cell

1. Pass cell suspension through cell strainer

2. Centrifuge at $400 \times g$ for 5 min

3. Resuspend cell pellet in 10 ml D10 containing 5% gentamycin sulfate

4. Plate cells in a 100-mm diameter culture dish and incubate overnight. Cells will begin to adhere within 24 h.

5. Remove medium after 48 h, carefully flush with PBS and replace with D10 containing 1% gentamycin sulfate.

6. Repeat Step 5, with change to D10 *without* gentamycin sulfate.

7. Colonies appear after 3 days. hfMSC should be subcultured at 70% confluence at 1:5 ratios (see Note 5).

3.3. Lentivector Preparation

Protocols and methods of lentiviral production, establishing virus titers and transduction are documented in this as well as in the previous edition of this book (see Note 6).

3.4. Transduction of hfMSC

1. Plate hfMSC at density of $4 \times 10^4/\text{cm}^2$ in 24-well plates.

2. Perform a single-round transduction in the presence of 4 µg/ml of polybrene.

3. Mock transduce hfMSC with addition of polybrene in D10 without vector.

4. Replace medium after 12 h.

5. Cells can be visualized under fluorescent microscope if reporter genes such as GFP have been used (see Note 7).

6. Flow cytometry can be used to perform sorting or analysis 72 h post-transduction; Immunostaining can be performed to detect phenotypical expression of transduced gene (see Fig. 1).

3.5. Immunostaining

1. Plate hfMSC at density of $4 \times 10^4/\text{cm}^2$ in chamber slides.

2. Immerse in ice-cold fixative and incubate for 15 min at room temperature.

3. Put in blocking agent at room temperature for 1 h (see Note 8).

4. Replace with primary antibody solution, incubate at room temperature for 1 h.

5. Wash twice with PBS, 5 min each wash.

6. Incubate with secondary antibody at room temperature for 1 h.

7. Wash twice PBS, 5 min each wash.

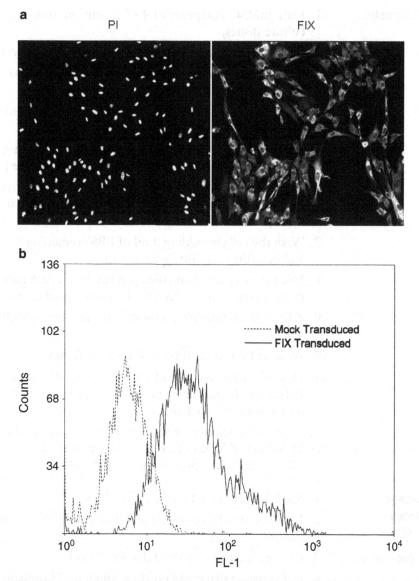

Fig. 1. Characterization of transduced cells. Lentiviral transduction of therapeutic genes can be achieved. Here, we characterize the transduced cells to verify expression of the gene of interest (**a**) Immunohistochemical analysis of Factor IX transduced cells. Transduced cells were fixed, permeabilised and stained with antibodies against human Factor IX (labeled with Alexafluo 488 antibody, *green*). Cell nuclei were counterstained with Proidium Iodide (*red*) Note the cytosolic staining pattern. Factor IX is a secretory protein stored within intracellular pools. In contrast, mock transduced cells will exhibit dim or diffuse, non-specific staining. (**b**) Flow cytometric analysis can be carried out to quantitatively assess positively transduced cells. Intracellular staining is carried out on a representative sample of transduced cells (*red*) and mock-transduced cells (*blue*). Note the peak shift indicating positive staining

8. Dip in 100% ethanol, air dry.

9. Apply mounting medium and place cover slip.

10. Image with epifluorescence or confocal microscope.

3.6. Flow Cytometric Analysis

1. Plate hfMSC at density of $4 \times 10^4/cm^2$ in 100-mm diameter culture dishes.

2. Culture to 90% confluence. This will yield 2×10^6 cells.

3. Lift cells by flushing lightly with PBS and incubating with 1 ml of trypsin for 5 min.

4. Remove cells, add D10 to block trypsin and centrifuge at $400 \times g$ for 10 min.

5. Resuspend in 100 µl 0.01% formaldehyde for 10 min at room temperature. This will fix the cells and stabilize the proteins.

6. Add 100 µl detergent, and incubate for 15 min at room temperature. This will permeabilise the cell membrane and reduce non-specific fluorescence.

7. Wash the cells by adding 2 ml of PBS containing 0.1% Triton and centrifuge at $300 \times g$ for 5 min.

8. Discard supernatant and resuspend in 90 µl FACS buffer $+10$ µl normal goat serum. This will block non-specific binding.

9. Add 2 µl of primary antibody, and incubate for 30 min at 4°C.

10. Wash in PBS, centrifuge at $400 \times g$ for 5 min.

11. Discard supernatant and resuspend in 4% formaldehyde. Labeled cells are stable and can be kept for a few days before flow cytometric analysis.

12. Data can be acquired using a fluorescence-activated cell sorter. Detection of AlexaFluo 488 can be performed by excitation at 488 nm and detection of emission at 530 nm.

3.7. Tri-Lineage Differentiation (see Note 9)

3.7.1 Bone Differentiation

1. Seed the cells at $3.1 \times 10^3/cm^2$ in D10.

2. Let the cells adhere overnight. Change to bone differentiation media.

3. Change media every third day for 3 weeks.

4. Cells change shape and produce minerals. Phosphate salts can be detected with Von Kossa staining (see below).

3.7.2. Adipogenic Differentiation

1. Seed the cells at $2.1 \times 10^4/cm^2$ in D10.

2. Let the cells adhere overnight.

3. Change to fat differentiation induction medial; incubate for 3 days.

4. Change to supportive media; incubate for 3 days.

5. Repeat steps 3 and 4 three times (see Note 10).

6. Let the cells grow in supportive media for 1 week.

7. Cells produce lipid vacuoles that can be stained with Oil Red O.

3.7.3. Chondrogenic
Differentiation

1. Wash the cells in cartilage differentiation medium *without* TGF-β3 by centrifuging at $150 \times g$ for 5 min at room temperature. Discard supernatant.

2. Resuspend 2.5×10^5 cells/polystyrene tube in medium *without* TGF-β3 and centrifuge again at $150 \times g$ for 5 min. Discard supernatant.

3. Resuspend the cells in 0.5 ml chondrogenic medium.

4. Centrifuge the cells at $150 \times g$ for 5 min and *do not* discard the supernatant.

5. Loosen the cap and place in the incubator.

6. Carefully replace medium every 2–3 days. (see Notes 11 and 12).

7. Harvest the cells after 28 days.

8. Wash twice with PBS and fix in 10% formalin for 1 h. Embed in paraffin for sectioning (4-μm thick sections).

3.8. Histological Staining (see Fig. 2)

3.8.1. Von Kossa

1. Wash the cells *very gently* twice with PBS.

2. Fix with 10% formalin for 1 h. Wash twice with dH_2O (see Note 13).

3. Stain with 2% Silver nitrate in dH_2O (w/v) for 10 min in the dark (see Note 14).

4. Expose to bright light for 15 min (not more than 1 h).

5. Wash twice with dH_2O and air dry.

6. Add 1% pyrogallol acid in dH_2O remove excess silver nitrate.

7. Wash twice with dH_2O.

8. Add 15% sodium thiosulfate fix the silver nitrate.

9. Wash twice with dH_2O. Image.

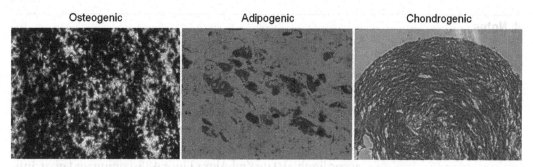

Osteogenic   Adipogenic   Chondrogenic

Fig. 2. Trilineage differentiation of fMSC and immunohistochemical staining. fMSC subject to osteogenic induction readily form bone, secreting calcium crystals, which are easily identified by von Kossa staining. Similarly, fMSC differentiated toward adipogenic lineage form lipid vesicles, identified by Oil Red O, and fMSC can be directed towards chondrogenic lineages, forming cartilaginous tissue, identified by Safranin O staining of proteoglycans

3.8.2. Oil Red O

1. Remove medium and wash cells *very gently* twice with PBS.
2. Fix with 10% formalin for 30 min at room temperature.
3. Carefully rinse cells with 60% isopropanol in H_2O.
4. Stain with Oil Red O for 10 min (see Note 15).
5. Wash four times with dH_2O. Image. For storage, keep immersed in dH_2O at 4°C.

   Following steps may be used to quantify staining

6. Remove all liquid. Add 100% isopropanol and pipette up and down a few times, to ensure all is Oil Red O extracted.
7. Transfer the supernatant to tubes.
8. Measure the optical density at 500 nm.
9. Compare the amount of dye in samples with a standard curve made with Oil Red O.

3.8.3. Safranin O

1. Deparaffinise and rehydrate sections through graduated washes (see Note 16):
   (a) Histoclear 2 min (two changes).
   (b) 100% EtOH 1 min (two changes).
   (c) 70% EtOH 30 s.
   (d) dH_2O 1 min.
2. Stain in 0.1% Safranin O solution for 5 min.
3. Wash twice in dH_2O.
4. Dehydrate in graded alcohol:
   (a) 70% EtOH for 30 s.
   (b) 100% EtOH for 30 s (two changes).
   (c) Clear in Histoclear.
5. Mount slides with DMX.

---

4. Notes

1. Vectors used in these experiments were a kind gift from Luigi Naldini, Didier Trono and Federico Marini.
2. Concentrations to be determined using suitable controls. 10 µg/ml were used in these experiments.
3. Suitable antibodies and concentrations to be determined accordingly. 10 µg/ml Alexa Fluor 488-conjugated goat antibodies were used in these experiments.
4. Sterile filter after addition of supplements.

5. Cells must be passaged before reaching confluence. Overly confluent cells will lose differentiation capacity.

6. Further information on lentivector production and titration can obtained from Trono Lab protocols (http://tronolab.epfl.ch/page58122.html)

7. Using a PGK.eGFP vector, a multiplicity of infection of 11 would provide a transduction efficiency of 98%.

8. Keep samples moist by incubating in a humid environment. Do not allow slides to dry off.

9. Toward the end of differentiation, cell sheets become loosely adhered to culture surface. Change the medium very gently.

10. Do not let the cells dry in between media changes.

11. Cell pellets may be loosely packed in the beginning; centrifuge the cells at $150 \times g$ for 5 min first before changing medium.

12. After each change of medium, tubes may be flicked to allow the cell pellet to float in the medium.

13. Avoid the use of phosphates-containing media at the stage to prevent false detection.

14. Silver nitrate should be freshly made.

15. Fresh Oil Red O working solution should be prepared at least 1 h before staining.

16. If the dH_2O goes cloudy at the end of washes, return the slides to 70% EtOH and take the slides through the ethanols again. Drain the slides from one dish before transferring them into the next dish to avoid too much carry-over.

References

1. Friedenstein, A.J., Deriglasova, U.F., Kulagina, N.N., Panasuk, A.F., Rudakowa, S.F., Luria, E.A., and Ruadkow , I.A. (1974) Precursors for fibroblasts in different populations of hematopoietic cells as detected by the in vitro colony assay method. *Exp Hematol* **2**, 83–92

2. Erices, A., Conget, P., and Minguell J.J. (2000) Mesenchymal progenitor cells in human umbilical cord blood. *Br J Haematol* **109**, 235–42

3. Plaas, H.A., and Cryer, A. (1980) The isolation and characterization of a proposed adipocyte precursor cell type from bovine subcutaneous white adipose tissue. *J Dev Physiol* **2**, 275–89

4. Romanov, Y.A., Svintsitskaya, V.A., and Smirnov, V.N. (2003) Searching for alternative sources of postnatal human mesenchymal stem cells: candidate MSC-like cells from umbilical cord. *Stem Cells* **21**, 105–10

5. Suchanek, J., Soukup, T., Ivancakova, R., Karbanova, J., Hubkova, V., Pytlik, R., and Kucerova, L. (2007) Human dental pulp stem cells – isolation and long term cultivation. *Acta Medica (Hradec Kralove)* **50**, 195–201

6. Tsai, M.S., Lee, J.L., Chang, Y.J., and Hwang, S.M. (2004) Isolation of human multipotent mesenchymal stem cells from second-trimester amniotic fluid using a novel two-stage culture protocol. *Hum Reprod* **19**, 1450–6

7. Campagnoli, C., Roberts, I.A.G., Kumar, S., Bennett, P.R., Bellantuono, I., and Fisk, N.M. (2001) Identification of mesenchymal stem/progenitor cells in human first-trimester fetal blood, liver, and bone marrow. *Blood* **98**, 2396–402

8. Chan, J., O'Donoghue, K., Gavina, M., Torrente, Y., Kennea, N., Mehmet, H., Stewart, H., Watt, D.J., Morgan, J.E., and

Fisk N.M. (2006) Galectin-1 induces skeletal muscle differentiation in human fetal mesenchymal stem cells and increases muscle regeneration. *Stem Cells* **24**, 1879–91

9. Guillot, P.V., Gotherstrom, C., Chan, J., Kurata, H., and Fisk, N.M. (2007) Human first-trimester fetal MSC express pluripotency markers and grow faster and have longer telomeres than adult MSC. *Stem Cells* **25**, 646–54

10. Kennea, N., Waddington, S., Chan, J., O'Donoghue, K., Yeung, D., Taylor, D., Al-Allaf, F., Pirianov, G., Themis, M., Edwards, D., Fisk, N., and Mehmet, H. (2009) Differentiation of human fetal mesenchymal stem cells into cells with an oligodedrocyte phenotype. *Cell Cycle* **8**, 1069–79

11. Gotherstrom, C., Ringden, O., Tammik, C., Zetterberg, E., Westgren, M., and Le Blanc, K. (2004) Immunologic properties of human fetal mesenchymal stem cells. *Am J Obstet Gynecol* **190**, 239–45

12. Gotherstrom, C., Ringden, O., Westgren, M., Tammik, C., and Le Blanc, K. (2003) Immunomodulatory effects of human foetal liver-derived mesenchymal stem cells. *Bone Marrow Transplant* **32**, 265–72

13. Zhang, Z.Y., Teoh, S.H., Chong, M.S., Schantz, J.T., Fisk, N.M., Choolani, M.A., and Chan, J. (2008) Superior osteogenic capacity for bone tissue engineering of fetal compared to perinatal and adult mesenchymal stem cells. *Stem Cells* **27**, 126–37

14. Le Blanc, K., Gotherstrom, C., Ringden, O., Hassan, M., McMahon, R., Horwitz, E., Anneren, G., Axelsson, O., Nunn, J., Ewald, U., Norden-Lindeberg, S., Jansson, M., Dalton, A., Astrom, E., and Westgren, M. (2005) Fetal mesenchymal stem-cell engraftment in bone after in utero transplantation in a patient with severe osteogenesis imperfecta. *Transplantation* **79**, 1607–14

15. Chan, J., Kumar, S., and Fisk, N.M. (2008) First trimester embryo-fetoscopic and ultrasound-guided fetal blood sampling for ex vivo viral transduction of cultured human fetal mesenchymal stem cells. *Hum Reprod* **23**, 2427–37

16. Guillot, P.V., Abass, O., Bassett, J.H., Shefelbine, S.J., Bou-Gharios, G., Chan, J., Kurata, H., Williams, G.R., Polak, J., and Fisk N.M. (2008) Intrauterine transplantation of human fetal mesenchymal stem cells from first-trimester blood repairs bone and reduces fractures in osteogenesis imperfecta mice. *Blood* **111**, 1717–25

17. Bartholomew, A., Patil, S., Mackay, A., Nelson, M., Buyaner, D., Hardy, W., Mosca, J., Sturgeon, C., Siatskas, M., Mahmud, N., Ferrer, K., Deans, R., Moseley, A., Hoffman, R., and Devine, S.M. (2001) Baboon mesenchymal stem cells can be genetically modified to secrete human erythropoietin in vivo. *Hum Gene Ther* **12**, 1527–41

18. Devine, S.M. (2002) Mesenchymal stem cells: will they have a role in the clinic? *J Cell Biochem* **85**, 73–9

19. Chan, J., Waddington, S.N., O'Donoghue, K., Kurata, H., Guillot, P.V., Gotherstrom, C., Themis, M., Morgan, J.E., and Fisk, N.M. (2007) Widespread distribution and muscle differentiation of human fetal mesenchymal stem cells after intrauterine transplantation in dystrophic mdx mouse. *Stem Cells* **25**, 875–84

20. Meyerrose, T.E., Roberts, M., Ohlemiller, K.K., Vogler, C.A., Wirthlin, L., Nolta, J.A., and Sands, M.S. (2008) Lentiviral-transduced human mesenchymal stem cells persistently express therapeutic levels of enzyme in a xenotransplantation model of human disease. *Stem Cells* **26**, 1713–22

21. Meyerrose, T.E., De Ugarte, D.A., Hofling, A.A., Herrbrich, P.E., Cordonnier, T.D., Shultz, L.D., Eagon, J.C., Wirthlin, L., Sands, M.S., Hedrick, M.A., and Nolta J.A. (2007) In vivo distribution of human adipose-derived mesenchymal stem cells in novel xenotransplantation models. *Stem Cells* **25**, 220–7

22. Kumar, S., Chanda, D., and Ponnazhagan, S. (2008) Therapeutic potential of genetically modified mesenchymal stem cells. *Gene Ther* **15**, 711–5

23. Reiser, J., Zhang, X.Y., Hemenway, C.S., Mondal, D., Pradhan, L., and La Russa, V.F. (2005) Potential of mesenchymal stem cells in gene therapy approaches for inherited and acquired diseases. *Expert Opin Biol Ther* **5**, 1571–84

24. Flake, A.W. (2004) In utero stem cell transplantation. *Best Pract Res Clin Obstet Gynaecol* **18**, 941–58

25. Muench, M.O. (2005) In utero transplantation: baby steps towards an effective therapy. *Bone Marrow Transpl* **35**, 537–47

26. Muench, M.O., Pott Bartsch, E.M., Chen, J.C., Lopoo, J.B., and Barcena, A. (2003) Ontogenic changes in CD95 expression on human leukocytes: prevalence of T-cells expressing activation markers and identification of CD95-CD45RO+ T-cells in the fetus. *Dev Comp Immunol* **27**, 899–914

27. Muench, M.O., Rae, J., Barcena, A., Leemhuis, T., Farrell, J., Humeau, L., Maxwell-Wiggins, J.R., Capper, J., Mychaliska, G.B., Albanese, C.T., Martin, T., Tsukamoto, A., Curnutte, J.T., and Harrison, M.R. (2001) Transplantation of a fetus with paternal Thy-1(+)CD34(+)cells for chronic granulomatous disease. *Bone Marrow Transpl* **27**, 355–64

28. Goodnow, C.C. (1996) Balancing immunity and tolerance: deleting and tuning lymphocyte repertoires. *Proc Natl Acad Sci U S A* **93**, 2264–71

29. Goodnow, C.C., Sprent, J., Fazekas de St Groth, B., and Vinuesa, C.G. (2005) Cellular and genetic mechanisms of self tolerance and autoimmunity. *Nature* **435**, 590–7

30. Amado, R.G., and Chen, I.S. (1999) Lentiviral vectors – the promise of gene therapy within reach? *Science* **285**, 674–6

31. Chan, J., O'Donoghue, K., de la Fuente, J., Roberts, I.A., Kumar, S., Morgan, J.E., and Fisk, N.M. (2005) Human fetal mesenchymal stem cells as vehicles for gene delivery. *Stem Cells* **23**, 93–102

Chapter 10

Manipulating the Cell Differentiation Through Lentiviral Vectors

Valeria Coppola, Cesare Galli, Maria Musumeci, and Désirée Bonci

Abstract

The manipulation of cell differentiation is important to create new sources for the treatment of degenerative diseases or solve cell depletion after aggressive therapy against cancer. In this chapter, the use of a tissue-specific promoter lentiviral vector to obtain a myocardial pure lineage from murine embryonic stem cells (mES) is described in detail. Since the cardiac isoform of troponin I gene product is not expressed in skeletal or other muscle types, short mouse cardiac troponin proximal promoter is used to drive reporter genes. Cells are infected simultaneously with two lentiviral vectors, the first expressing EGFP to monitor the transduction efficiency, and the other expressing a puromycin resistance gene to select the specific cells of interest. This technical approach describes a method to obtain a pure cardiomyocyte population and can be applied to other lineages of interest.

Key words: Murine embryonic stem cells, Lentiviral vectors, TNNI promoter, Myocardial pure lineage, EGFP

1. Introduction

Manipulating cell differentiation offers the perspective to create new sources for the treatment of degenerative diseases or solve cell depletion after aggressive therapy against cancer. In the last years, the efforts of the scientific community have been focused on finding pluripotent stem cell sources by optimizing the expansion and manipulation of embryonic stem cells and evaluating the multipotency of adult stem cell subpopulations (1–3). A new emerging field is the reprogramming of somatic cells to an embryonic-like state. In this regard, recent papers report that differentiated cells such as fibroblasts can generate induced pluripotent stem cells (iPS) after high efficiency retrovirus-mediated transduction

Maurizio Federico (ed.), *Lentivirus Gene Engineering Protocols,* Methods in Molecular Biology, vol. 614,
DOI 10.1007/978-1-60761-533-0_10, © Humana Press, a part of Springer Science+Business Media, LLC 2010

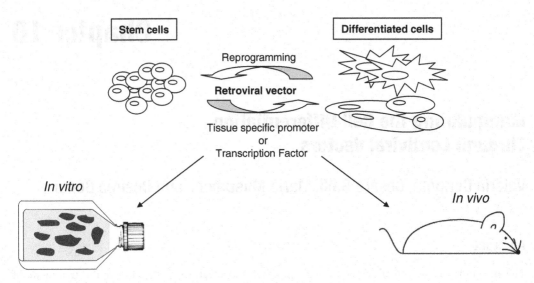

Fig. 1. General scheme of the possible applications of retroviral vectors

of four factors (e.g., OCT-4, Sox2, Kfl-4, Myc) (Fig. 1) (4–8). Retroviral particles transducing transcription factors critical for the induction of specific differentiation (e.g., GATA4, Twist) or signaling molecules (e.g., insulin-like growth factor II, Cripto) can be used to obtain pure lineages of cells of interest (9). Moreover, the application of vector technology gives a great input to isolate subpopulations using specific promoters or genomic loci (e.g., MLC2v, cTnI, SP-C, CK6, Nanog, Fbxo15) that selectively drive reporter genes (6, 10, 11). Furthermore, lentiviral transgene delivery offers a spectacularly efficient method for the generation of large transgenic animals through the infection of oocytes or embryos soon after fertilization (12, 13).

Embryonic stem cells (ES) harvested from the inner cell mass of the blastocyst can proliferate indefinitely in vitro while retaining the ability to differentiate into somatic cells (9, 14–20). These properties make ES an attractive tool for cell replacement therapies of degenerative disorders associated with a loss of functional cells, such as myocardial infarction, diabetes, and Parkinson's disease. In vitro, ES cells can be propagated in the undifferentiated state in the presence of leukemia inhibitory factor (LIF) or in coculture with mitotically inactivated mouse embryonic fibroblasts (MEF). In the absence of these conditions, the cells undergo a spontaneous differentiation program that reflects the critical stages of the embryo development. The most common method for initiating differentiation is the formation of embryoid bodies (EBs), spheroid structures growing in suspension culture and containing derivatives of all three embryonic germ layers. EBs return to adherent culture conditions in the presence of the appropriate stimuli, which allow the recovery of almost all cell lineages of body tissues.

The routinely used differentiation protocols give rise to a mixture of different cell populations. Obtaining pure populations remains one of the most important conditions to use stem cells as a possible therapeutic source (8, 21). For this end, genetic manipulation of ES cells by lentiviral vectors has proven to be the most effective and stable system of obtaining transgenic embryonic stem cell lines. Lentiviruses were shown to drive gene expression efficiently in various types of stem cells, with little immunogenic effect and less frequent silencing of the transgene than Moloney-derived retroviruses, thus offering an easy and effective tool to enhance the yield and the purity of differentiated cells.

The following paragraphs describe an experimental approach to obtain a pure myocardial lineage from mouse embryonic stem cells. Cells are infected simultaneously with two different lentivirus vectors, one expressing EGFP to monitor the transduction efficiency, and the other carrying the puromycin resistance gene to select the transduced cells. Short mouse cardiac troponin proximal promoter (22) has been used to drive both EGFP and puromycin resistance genes. The cardiac isoform of troponin I gene product (TcIc) is not expressed in skeletal or other muscle types. Human and mouse TnIc promoter regions have been studied previously in order to identify potential *cis*-acting DNA elements conferring cardiac specificity (23, 24). The phosphoglycerate kinase (PGK) regulatory region is used as control ubiquitous promoter in the reported method. This technique is conducted in mouse embryonic stem cells, but a recent paper describes similar methods applied on human ES (11).

2. Materials

2.1. Murine Embryonic Stem Cells

1. Feeder-independent murine embryonic stem cells (mES) lines (see Note 1).
2. Gelatine-coated plates (Invitrogen) (see Note 2).
3. Glasgow MEM/BHK medium containing: 10% fetal bovine serum (FBS, by Hyclone, Logan, UT), 0.23% sodium bicarbonate, 100× Non-Essential Amino Acids (NEAA), 1 nM sodium pyruvate, 100 nM 2-mercaptoethanol and 2 mM L-glutamine (see Note 3).
4. Leukemia-inhibitory factor (LIF, Chemicon International, Temecula, CA) (see Note 4).
5. 0.25% Trypsin/EDTA solution (GIBCO) with a pH-indicator (see Note 5).
6. Uncoated plastic such as bacteriological plates to produce embryoid bodies (EBs).
7. 0.1% dimethylsulfoxide (DMSO).

2.2. Lentiviral Vector with Tissue-Specific Promoter

Three-plasmid expression system to generate lentiviral vectors: (a) the packaging construct, pCMV R8.74, designed to provide the HIV-1 proteins needed to produce the lentiviral vector particle; (b) the envelope-coding plasmid, pMD.G, for pseudotyping the virion with the Vesicular Stomatitis Virus Glycoprotein (VSV-G), and (c) the following self-inactivating (SIN) transfer vector plasmids (see Fig. 2a) (see Note 6):

pRRLcPPT.hPGK.EGFP.WPRE	(indicated as pPGK.EGFP)
pRRLcPPT.hPGK.PURO.WPRE	(indicated as pPGK.PURO)
pRRLcPPT.mTNNI.EGFP.WPRE	(indicated as pTNNI.EGFP)
pRRLcPPT.mTNNI.PURO.WPRE	(indicated as pTNNI.PURO)

2.3. Lentiviral Vector Production

1. 293T cells (human embryonic kidney cell line also called 293tsA1609ne, derived from 293 cells).

2. IMDM supplemented with 10% FBS, L-glutamine (50 U/ml), and penicillin/streptomycin (50 U/ml).

3. 2× HBS (Hepes-Buffered Saline, Sigma): 280 mM NaCl, 10 mM KCL, 1.5 mM Na_2HPO_4, 12 mM d(+)glucose,

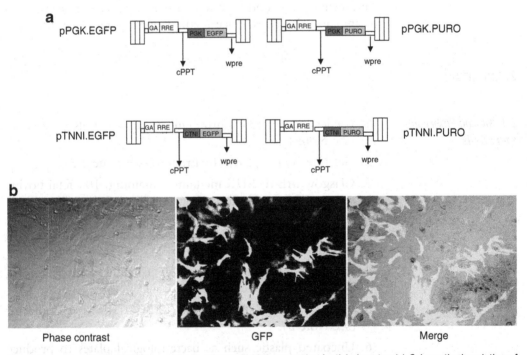

Fig. 2. Tissue specificity of TNNI promoter in advanced third generation lentiviral vector. (**a**) Schematic description of lentiviral vectors pRRLcPPT.hPGK.EGFP.WPRE(pPGK.EGFP), pRRLcPPT.hPGK.PURO.WPRE(pPGK.PURO), pRRLcPPT.mTNNI.EGFP.WPRE(pTNNI.EGFP), pRRLcPPT.mTNNI.PURO.WPRE(pTNNI.PURO). (**b**) Specific expression of pTNNI.EGFP lentiviral vector in cardiomyocytes. *EGFP* enhanced green fluorescent protein

50 mM HEPES, pH 7.05. Store in aliquots at –20°C and utilize within 6 months. Wait till the solution reaches room temperature, before use.

4. Solution of 2 M $CaCl_2$ (125 mM, final concentration in DNA/HBS solution): filter the solution and store in aliquots at –20°C; H_2O: use cell culture-grade water.

5. DNA for transfection: use an endotoxin free DNA for transfection, extracted with the Endofree Plasmid Mega or Maxi Plasmid kit (Qiagen) (see Note 7).

6. Polybrene (Sigma): dissolve the powder in cell culture-grade water at a final concentration of 8 mg/ml. Make a 2,000× stock solution. Store in aliquots at –20°C.

7. CellQuest (Becton Dickinson) or WindMDI (Microsoft) softwares for the elaboration of FACS analysis of transduced cells.

2.4. Primary Cardiomyocytes

1. DMEM-Medium199 (4:1) supplemented with 5% FBS, 5% HS (Horse Serum), 1% l-glutamine and 1% penicillin/streptomycin (Invitrogen, Carlsbad, CA).

2. Mitomycin C (Sigma) (see Note 8).

3. Methods

3.1. Recovery of Lentiviral Vectors

1. Transfect 293T cells with packaging/lentiviral vector/VSV-G expressing constructs with the most familiar method. We use to transfect cells with the broadly described calcium phosphate method.

2. Collect the packaging cell supernatant 48 h after transfection, centrifuge it at $180 \times g$ for 5 min to clear it of cell debris, and filter through 0.45 μm sized filter. Aliquot viral supernatants in polypropylene tubes and freeze them in liquid nitrogen. Store at –80°C.

3. If you need concentrated viral particles, ultracentrifuge medium at $50,000 \times g$ for 90 min at 4°C. Resuspend pellets with PBS containing 0.5% BSA (see Note 9), and store in liquid nitrogen. Use polyallomer ultracentrifuge tubes.

3.2. mES Culture

1. Cultivate mES cells in gelatine-coated plates in cultivation medium indicated in materials. The medium must be changed every 2 days.

2. Split the cells once or twice a week evaluating cell aspect: the medium must not become too acid (yellow), but cells must not be cultivated too diluted. It is possible to evaluate as optimal density 50–60% of confluence in a 100-mm plate, corresponding to about 4×10^6 cells/plate (see Notes 10 and 11).

3. To initiate differentiation, detach mES cells, disperse them in single cells, and seed in uncoated plastic (i.e., bacteriological plates) to produce embryoid bodies.

4. When EBs are formed, replace medium and maintain embryoid bodies for 3 days in cultivation medium supplemented with 0.1% dimethylsulfoxide (DMSO) but without LIF. Maintain EBs in suspension 2 more days with cultivation medium without LIF or DMSO, then allow them to settle onto gelatine-coated plates in the presence of cultivation medium without LIF or DMSO. Change medium every 2 days.

5. Observe the culture under the microscope, beating areas will appear within 16–18 days.

3.3. Transduction of mES with Lentivirus Particles

1. Plate cells in 100 mm dishes at final confluence of 40–50% the day before the infection (about $3-4 \times 10^6$ cells/plate).

2. On the day of infection, remove the medium and replace with lentiviral vector containing supernatant in the presence of 4 µg/ml of Polybrene. Centrifuge cells for 45 min at $600 \times g$ at 32°C, keep for 75 min in a 5% CO2 incubator at 37°C, then replace medium.

3. Infect mES cells with different lentiviral vectors produced by vectors pPGK.EGFP, pPGK.PURO, pTNNI.EGFP, pTNNI.PURO.

4. Perform preliminary experiments using different amounts of lentiviral vector and individuate the dose infecting 100% of the population without toxicity. Since in the final experiment mES cells must be infected simultaneously with two different lentiviral vectors at the 1:1 ratio (i.e., pPGK.EGFP/pPGK.PURO or pTNNI.EGFP/pTNNI.PURO), make sure that half of the dose gives 100% of population infected as well (Fig. 3a). The suggested doses are 1×10^6 TU/ml and 5×10^5 TU/ml in the presence of Polybrene (TU/ml: number of Transducing Units per milliliter of viral supernatant) (see Note 12).

5. 48 h after transduction, analyze EGFP infected samples by FACS and test puromycin resistance. The totality of cells is expected to be infected, meanwhile showing antibiotic resistance. Evaluate cell death with 7-Aminoactinomycin D (7AAD) staining. (Figs. 3b, c and 4).

3.4. Transduction of Cardiomyocytes with Lentivirus Particles

1. For the preparation of primary cardiomyocytes, please refer to the previous Edition of this book.

2. Plate 2.5×10^4 of primary cardiomyocytes 2 days before exposure to the virus in 24-well plate.

3. On the day of infection, remove the medium, and replace with viral supernatant in the presence of 4 µg/ml of Polybrene.

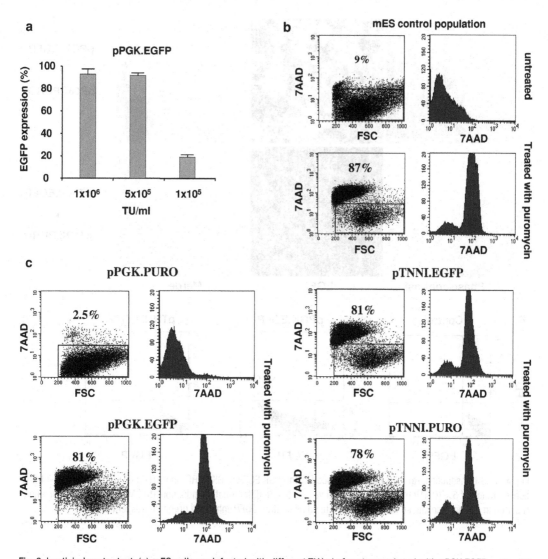

Fig. 3. Lentiviral vector test. (**a**) mES cells are infected with different TU/ml of vector produced with pPGK.EGFP vector. (**b**) mES cells treated or untreated with 1 µg/ml of puromycin and analyzed for 7AAD staining by FACS. (**c**) mES cells infected with 5 × 10⁵TU/ml of pPGK.EGFP or pPGK.PURO and pTNNI.EGFP or pTNNI.PURO, treated with 1 µg/ml of puromycin and analyzed for 7AAD staining by FACS

Centrifuge cells for 45 min at $600 \times g$ at 32°C, then keep for 75 min in a 5% CO_2 incubator at 37°C.

4. Replace medium and analyze the levels of cell transduction by FACS after 48 h.

3.5. mES Differentiation

After the transduction with mixed lentiviral vector preparations produced by vectors pPGK.EGFP/pPGK.PURO or pTNNI.EGFP/pTNNI.PURO, cell differentiation has to be induced as follows:

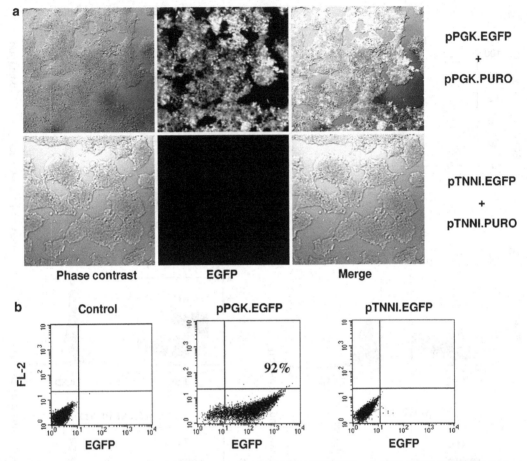

Fig. 4. mES transduction. (**a**) mES cells transduced with pPGK.EGFP/pPGK.PURO and pTNNI.EGFP/pTNNI.PURO (at final concentration of 5 × 10⁵ TU/ml for each lentiviral vector) before differentiation induction. (**b**) FACS analysis of populations reported in (**a**); not transduced mES have been used as control. *EGFP* enhanced green fluorescent protein

1. 48 h after transduction, shortly trypsinize cells to eliminate clumps, and plate them for EBs formation. Plate in a 100 mm dish of untreated sterile plastic diluting the original culture 1:6, corresponding to about 0.5–1 × 10⁶ cells. Maintain the cultivation medium without removing LIF.

2. After 2 days, add 0.1% DMSO to the medium, remove LIF and change the medium once a day for 3 days (see Note 13).

3. Wash EBs with PBS and maintain them for 2 more days in cultivation medium without LIF and DMSO.

4. Collect the EBs, wash them in PBS and plate them again counting about 20–50 EBs per 35 mm gelatine-coated dish. Continue cultivation without LIF and DMSO.

5. Change medium every 2 days.

pPGK.EGFP/pPGK.PURO   pTNNI.EGFP/pTNNI.PURO

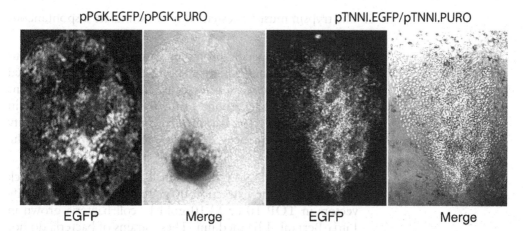

EGFP   Merge   EGFP   Merge

Fig. 5. Myocardial beating areas visible in cultures at 16–18 days. EGFP expression driven by TNNI promoter reveals myocardial differentiation (*right panels*). *EGFP* enhanced green fluorescent protein. Merge: phase contrast and EGFP

3.6. Monitoring mES Differentiation

1. After 7–9 days from EBs plating on gelatine, observe beating areas of cells under the microscope.

2. At this stage, within myocardial cells it is possible to observe well differentiated beating EGFP positive areas under the microscope.

3. Put culture in selection with puromycin to increase the percentage of unilineage specific cells that should reach a quite pure myocardial population (see Notes 14 and 15) (Fig. 5). Add 1 μg/ml of puromycin in the culture medium and repeat the treatment changing medium every day until total death of the contaminating cells.

4. Evaluate also the control populations, such as pPGK.EGFP/ pPGK.PURO and pPGK.EGFP transduced cells, which represent the experiment controls (see Note 16).

4. Notes

1. Clones of these cells can be obtained upon request from the authors of the papers referenced in (25, 26).

2. Gelatine can be stored at 4°C for not more than 1 week. It can induce toxicity and differentiation, so it is suggested to test different batches directly on an aliquot of mES cells before initiating the cultivation.

3. FBS should be free of endotoxin.

4. The cultivation medium must be prepared fresh every week, especially if it includes LIF.

5. The trypsin must be tested, as it can stimulate a spontaneous differentiation.

6. When a tissue specific promoter is moved from a vector to another, the tissue or lineage specificity must be reconfirmed by infecting lineage-specific cells. It is also very important to evaluate the background of the tissue-specific promoter in any subcloning vector. Background could be due to the activity of enhancer sequences or other promoters present in the subcloning vector.

7. The impurity of DNA reduces transfection efficiency, so check the quality of the extraction by gel electrophoresis. Amplify vectors in TOP 10 or XL10 gold E. coli bacteria grown in Luria-Bertani (LB) medium. These strains of bacteria do not recombine plasmids.

8. Treat cells with Mitomycin C (Sigma) to prevent fibroblast cell growth. Final cell populations must contain more than 95% growth-arrested cardiomyocytes.

9. The pellets are not observable to the naked eye and are difficult to resuspend. Therefore, they should be resuspended with prolonged pipetting (use filtered tips) and then kept under the hood for at least 15 min before harvesting.

10. It is important that the mES retain their capacity to differentiate in different lineages, therefore it is crucial to control the quality of initial mES culture and to maintain the cell population in an undifferentiated state. To this end, it is useful to test the capacity of mES to differentiate into different tissues, every month (7, 9, 11–13, 20). In addition, the evaluation of OCT-4 gene expression could be a good indication of stem cell phenotype.

11. Freezing and thawing mES cells are delicate passages that can impact on cell differentiation. Be sure to freeze in FBS with 10% DMSO, and that, after thawing, cells retain the initial aspect and differentiation capacity. Freeze cells at a final concentration $>5 \times 10^6$ cells/vial.

12. The suggested doses are 10^6 TU/ml and 5×10^5 TU/ml, but they can change according to cell viability and lentiviral vector preparation. The collected supernatants have a concentration of about 1×10^6 TU/ml.

13. During centrifugation of EBs, check the speed. It must not be more than $100 \times g$ to avoid aggregation. Alternatively, leave the EBs to spontaneously sediment at the bottom of a 15 ml tube.

14. The obtained pure population can be controlled by immunostaining against mouse cardiac troponin I.

15. This is applicable to every kind of differentiation and tissue specific promoter.

16. In pPGK.EGFP/pPGK.PURO cell population, EGFP is visible and all cells are resistant to puromycin treatment; the pPGK.EGFP population is EGFP positive, but dies after puromycin treatment.

Acknowledgments

The authors thank Antonio Addario e Giuseppe Loreto for their professional technical support. The authors also thank Antonia Follenzi and Luigi Naldini for providing lentiviral vectors. This work was supported by the Italian Health Ministry and Istituto Superiore di Sanità.

References

1. Conrad S, Renninger M, Hennenlotter J, Wiesner T, Just L, Bonin M, Aicher W, Buhring HJ, Mattheus U, Mack A, Wagner HJ, Minger S, Matzkies M, Reppel M, Hescheler J, Sievert KD, Stenzl A, and Skutella T. (2008) Generation of pluripotent stem cells from adult human testis. *Nature* **456**, 344–9.

2. Galli C, Lagutina I, Crotti G, Colleoni S, Turini P, Ponderato N, Duchi R, and Lazzari G. (2003) Pregnancy: a cloned horse born to its dam twin. *Nature* **424**, 635.

3. Passier R, van Laake LW, and Mummery CL. (2008) Stem-cell-based therapy and lessons from the heart. *Nature* **453**, 322–29.

4. Jaenisch R, and Young R. (2008) Stem cells, the molecular circuitry of pluripotency and nuclear reprogramming. *Cell* **132**, 567–82.

5. Kim JB, Zaehres H, Wu G, Gentile L, Ko K, Sebastiano V, Arauzo-Bravo MJ, Ruau D, Han DW, Zenke M, and Scholer HR. (2008) Pluripotent stem cells induced from adult neural stem cells by reprogramming with two factors. *Nature* **454**, 646–50.

6. Nakagawa M, Koyanagi M, Tanabe K, Takahashi K, Ichisaka T, Aoi T, Okita K, Mochiduki Y, Takizawa N, and Yamanaka S. (2008) Generation of induced pluripotent stem cells without Myc from mouse and human fibroblasts. *Nat Biotechnol* **26**, 101–6.

7. Okita K, Nakagawa M, Hyenjong H, Ichisaka T, and Yamanaka S. (2008) Generation of mouse induced pluripotent stem cells without viral vectors. *Science* **322**, 949–53.

8. Takahashi K, Okita K, Nakagawa M, and Yamanaka S. (2007) Induction of pluripotent stem cells from fibroblast cultures. *Nat Protoc* **2**, 3081–89.

9. Wobus AM, and Boheler KR. (2005) Embryonic stem cells: prospects for developmental biology and cell therapy. *Physiol Rev* **85**, 635–78.

10. Brunetti D, Perota A, Lagutina I, Colleoni S, Duchi R, Calabrese F, Seveso M, Cozzi E, Lazzari G, Lucchini F, and Galli C. (2008) Transgene expression of green fluorescent protein and germ line transmission in cloned pigs derived from in vitro transfected adult fibroblasts. *Cloning Stem Cells* **10**, 409–19.

11. Gallo P, Grimaldi S, Latronico MV, Bonci D, Pagliuca A, Gallo P, Ausoni S, Peschle C, and Condorelli G. (2008) A lentiviral vector with a short troponin-I promoter for tracking cardiomyocyte differentiation of human embryonic stem cells. *Gene Ther* **15**, 161–70.

12. Chan AW, Homan EJ, Ballou LU, Burns JC, and Bremel RD. (1998) Transgenic cattle produced by reverse-transcribed gene transfer in oocytes. *Proc Natl Acad Sci U S A* **95**, 14028–33.

13. Hofmann A, Kessler B, Ewerling S, Weppert M, Vogg B, Ludwig H, Stojkovic M, Boelhauve M, Brem G, Wolf E, Pfeifer A. (2003) Efficient transgenesis in farm animals by lentiviral vectors. *EMBO Rep* **4**, 1054–60.

14. Dani C, Smith AG, Dessolin S, Leroy P, Staccini L, Villageois P, Darimont C, and Ailhaud G. (1997) Differentiation of embryonic stem cells into adipocytes in vitro. *J Cell Sci* **110**, 1279–85.

15. Doss MX, Koehler CI, Gissel C, Hescheler J, and Sachinidis A. (2004) Embryonic stem cells: a promising tool for cell replacement therapy. *J Cell Mol Med* **8**, 465–73.

16. Kawamorita M, Suzuki C, Saito G, Sato T, and Sato K. (2002) In vitro differentiation of

mouse embryonic stem cells after activation by retinoic acid. *Hum Cell* **15**, 178–82.

17. Rippon HJ, and Bishop AE. (2004) Embryonic stem cells. *Cell Prolif* **37**, 23–34.

18. Rippon HJ, Lane S, Qin M, Ismail NS, Wilson MR, Takata M, Bishop AE. (2008) Embryonic stem cells as a source of pulmonary epithelium in vitro and in vivo. *Proc Am Thorac Soc* **5**, 717–22.

19. Tang S, Qiu G, and Huang B. (2000) Differentiation of embryonic stem cells into neuronal cells in vitro. *Zhonghua Yi Xue Za Zhi* **80**, 936–8.

20. Wei H, Juhasz O, Li J, Tarasova YS, and Boheler KR. (2005) Embryonic stem cells and cardiomyocyte differentiation: phenotypic and molecular analyses. *J Cell Mol Med* **9**, 804–17.

21. Narazaki G, Uosaki H, Teranishi M, Okita K, Kim B, Matsuoka S, Yamanaka S, and Yamashita JK. (2008) Directed and systematic differentiation of cardiovascular cells from mouse induced pluripotent stem cells. *Circulation* **118**, 498–506.

22. Bhavsar PK, Brand NJ, Yacoub MH, and Barton PJ. (1996) Isolation and characterization of the human cardiac troponin I gene (TNNI3). *Genomics* **35**, 11–23.

23. Ausoni S, Campione M, Picard A, Moretti P, Vitadello M, De Nardi C, and Schiaffino S. (1994) Structure and regulation of the mouse cardiac troponin I gene. *J Biol Chem* **269**, 339–46.

24. Bonci D, Cittadini A, Latronico MV, Borello U, Aycock JK, Drusco A, Innocenzi A, Follenzi A, Lavitrano M, Monti MG, Ross J, Jr., Naldini L, Peschle C, Cossu G, and Condorelli G. (2003) 'Advanced' generation lentiviruses as efficient vectors for cardiomyocyte gene transduction in vitro and in vivo. *Gene Ther* **10**, 630–6.

25. Puceat M. (2008) Protocols for cardiac differentiation of embryonic stem cells. *Methods* **45**, 168-71.

26. Savatier P, Lapillonne H, Jirmanova L, Vitelli L, and Samarut J. (2002) Analysis of the cell cycle in mouse embryonic stem cells. *Methods Mol Biol* **185**, 27-33.

Chapter 11

Lentiviral Vector Transduction of Dendritic Cells for Novel Vaccine Strategies

Lung-Ji Chang

Abstract

Dendritic cells (DCs) are key antigen-presenting cells that induce primary and memory immune response. Patients with chronic infections or cancer often display DC dysfunctions. Modification of DCs or DC progenitors in vitro may overcome the problems with defective DCs in vivo. Lentiviral vector is highly efficient in transducing hematopoietic cells including DCs. Examples of lentiviral modification of DCs with immune modulatory genes and analysis of antigen-specific T cells to demonstrate enhanced immune effector functions of DCs will be introduced.

Key words: Lentiviral vector, Dendritic cell, T cells, Immunotherapy, Calnexin

1. Introduction

DCs are important immune cells in initiating and maintaining immunity (1). Thousands of patients have entered clinical trials employing DCs as therapeutic vaccines, yet only limited success has been achieved (2). The reason is that most patients have experienced a high burden of pathogenic antigens or cancer antigens. These patients' immune system would have been highly tolerized for the disease antigens, and may also display defective DC functions. Modification of DCs or DC precursors may overcome this problem (3–6).

Various approaches have been attempted to modify DC functions, which involve the use of inactive inflammatory bacterial products, cytokines, chemokines, Toll-like receptor ligands, and ligands of signaling pathways known to activate immune response (7, 8). Nevertheless, these exogenous factors may not effectively correct the intrinsic DC dysfunction. Intracellular modification of

Maurizio Federico (ed.), *Lentivirus Gene Engineering Protocols*, Methods in Molecular Biology, vol. 614,
DOI 10.1007/978-1-60761-533-0_11, © Humana Press, a part of Springer Science+Business Media, LLC 2010

DCs or DC precursors with genetic approaches, on the other hand, may overcome the intrinsic defects of DCs (9–12).

Genetically modified DCs can potentially activate a strong protective immunity against infections and cancer (12–16). Overexpression of an antigen without immune modulation in DCs may not be sufficient to induce a strong antigen-specific immune response. To establish an effective immune response, optimal interactions between antigen presenting cells and immune effector cells are required; these interactions are regulated by multiple cellular mediators including major histocompatibility complex (MHC) molecules (class I and class II) harboring antigenic peptides (signal 1), costimulatory molecules such as CD80, CD86, and ICAM-1 (signal 2), as well as cytokines, such as IL-2 and IL-12 (signal 3) (17). Up- or down-regulation of these immune modulatory genes in DCs may potentiate antigen-specific T cell responses (18, 19).

Efficient gene transfer into DCs has been difficult, but lentiviral vectors (LVs) transduce DCs at high efficiencies with little to no cytotoxicity (11, 20–22). For examples, lentiviral transduction of DCs to overexpress IL-12 or to express small interfering RNA (siRNA) targeting IL-10 has been shown to promote a strong Th1 response (11). Besides cytokine signaling, increased interaction of costimulatory molecules and adhesion molecules such as CD80, CD86, CD137, and ICAM-1 are also critical for the development of long-lasting memory immunity (23–26). Other immune modulatory approaches, including supraphysiological expression of a chaperone protein, calnexin, in DCs have also proven to be effective in overcoming the suppressive regulatory T cell activity of cancer patients (27).

The genetic constructs of an improved self-inactivating lentiviral vector system (NHP/TYF) expressing a reporter gene (placenta alkaline phosphatase, PLAP) and siRNA (driven by a polymerase III promoter) are illustrated in Fig. 1. The pTYF-PLAP lentiviral vector transduces human immature dendritic cells at very high efficiency (>90%, Fig. 2). To demonstrate that lentiviral modification of DCs can enhance the immune activation function of DCs, a chaperone protein calnexin was overexpressed in DCs with lentiviral vector, and the DCs were shown to substantially activate T cell expansion and induce a strong antigen-specific T cell response (Fig. 3).

2. Materials

2.1. Lentiviral Vectors and Transduction of DCs

1. An improved lentivirus vector system, NHP/TYF, is illustrated in Fig. 1. All lentiviral backbone plasmid components can be obtained from National Institutes of Health, Bethesda, MD, USA (see Note 1).

An Improved NHP/TYF Lentiviral Vector System

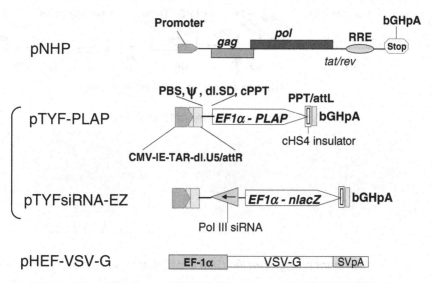

Fig. 1. Genetic constructs of the self-inactivating NHP/TYF lentiviral vector system. The NHP/TYF LV system uses three basic vector constructs, pNHP, pTYF and pHEF-VSVG. The entire 3′ U3 and U5 are deleted except for the 24 nt att site [32]

Lentiviral Transduction of Human Dendritic Cells

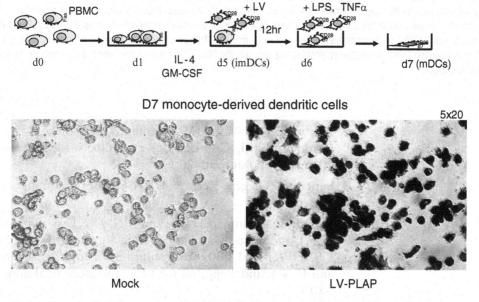

Fig. 2. Lentiviral transduction of human immature DCs. Human monocytes-derived dendritic cells (DC) were generated from healthy donors' peripheral mononuclear cells (PBMC) as illustrated. On day 5, the immature DC were infected with lentiviral vectors encoding PLAP or control vectors at 30–50 multiplicity of infection. After 2 days, the transduced DCs were stained for PLAP expression

DC Modification and Activation of Immune Response

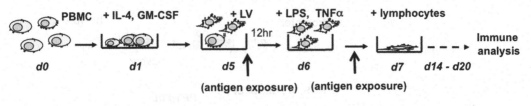

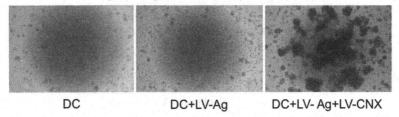

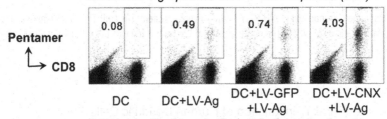

Fig. 3. Enhanced Ag-specific T cell response by LV-CNX modified DCs. DCs were transduced with LVs encoding a viral antigen (LV-Ag) together with LV-CNX (chaperone) to enhance antigen presentation function. Autologous immune cells were cocultured with the day-7 mature DCs for an additional 10–14 days. Substantially increased expansion of the cocultured immune cells in the presence of LV-CNX-DCs could be visualized in the 96-wells as illustrated. The day-20 antigen-specific CD8 T cells were then quantified with peptide-MHC pentamer 6 h after restimulation with specific antigen-presenting DCs. The pentamer-positive antigen-specific CD8 T cells were gated and the percentages indicated in the flow graph

2. Polybrene. Dissolve in AIM-V medium (Gibco-BRL, Invitrogen Corp., Carlsbad, CA, USA) as a 4,000× stock solution of 20 mg/ml and store at –20°C.

3. Superfect (Qiagen, Valencia, CA, USA) or other transfection reagents.

4. Recombinant human TNF α and IFN-γ R&D systems, MN, USA), and lipopolysaccharide (LPS, Sigma, St. Louis, MO, USA). Prepare 1,000× stock solutions (50 µg/ml for TNFα and IFN-γ, and 1 mg/ml for LPS) and store in aliquots at –80°C.

5. Sterile 0.45 mm low protein binding filters (Millipore, Billerica, MA, USA).

6. Glutaraldehyde fixation solution: 1% formaldehyde (0.27 ml of 37.6% stock to make 10 ml), 0.2% glutaraldehyde (80 µl of 25% stock to make 10 ml) in PBS, freshly prepared.

7. Alkaline phosphatase substrate kit, from Vector Laboratories Inc. Burlingame, CA, USA (BCIP/NBT, Cat.# SK-5400).

2.2. Preparation of Peripheral Blood Mononuclear Cells (PBMCs) and DCs

1. Ficoll–Hypaque solution (density 1.077 g/L), GE Healthcare Bio-Sciences (Buckinghamshire, UK).

2. Complete RPMI 1640 medium (Gibco-BRL) supplemented with 10% 56°C/30 min heat-inactivated fetal bovine serum (FBS) (HyClone, Logan, UT, USA), 100 µg/ml streptomycin and 100 IU/ml penicillin (Gibco-BRL).

3. AIM-V medium (Gibco-BRL).

4. Recombinant human GM-CSF and IL-4 (Biosource, CA, USA). Prepare 1,000× stock (50 µg/ml of GM-CSF and 25 µg/ml of IL-4) and store in aliquots at –80°C.

5. 6-well and 24-well low attachment plates (Corning-Costar, NY, USA).

2.3. Flow Cytometry Analysis

1. FACS buffer: PBS containing 2% FBS and 0.09% Na_3N.

2. Fixation solution: FACS buffer containing 1% formaldehyde (Fisher Biotech, NJ).

3. Flow cytometry tubes (BD Biosciences, San Diego, CA, USA).

4. Normal mouse serum (Sigma) for blocking non-specific antibody (Ab) binding.

5. Fluorochrome-labeled anti-human CD8 Abs and isotype controls (BD Biosciences).

2.4. Activation of Antigen-Specific T Cells by LV-Modified DCs

1. 96-well U-shape plates (Costar).

2. Recombinant human IL-2, IL-7 and IL-15 (Gentaur, Aachen, Germany). Prepare IL-2 (12.5 U/ml), IL-7 (10 ng/ml) and IL-15 (20 ng/ml) in sterile PBS with 0.1% BSA and store in aliquots at –80°C.

3. Anti-CD3 Ab for T cell activation as positive control (eBioscience, CA, USA)

2.5. MHC-Peptide Multimer Analysis

1. PE- or APC-conjugated MHC-peptide specific tetramers (Beckman Coulter, CA, USA) or pentamers (Proimmune, Bradenton, FL, USA). Alternatively, they can be obtained from MHC Tetramer Core Facility, National Institutes of Health (http://www.niaid.nih.gov/reposit/tetramer/genguide.html).

3. Methods

3.1. Preparation of Lentiviral Vectors

1. Prepare high purity of vector plasmids including packaging construct (pNHP), envelope construct (pHEF-VSV-G), and transducing construct (pTYF or pTYcPPT) for DNA cotransfection (see Note 1). Figure 11.1 illustrates lentiviral vectors containing a polymerase II promoter driven gene and polymerase III promoter driven siRNA gene.

2. Split 293T cells using trypsin and plate into 6-well plates at 90% confluency 16–24 h prior to DNA co-transfection.

3. Transfect 293T cells with the four plasmid DNAs for vector production using Superfect kit following the manufacturer's instructions.

4. Harvest virus supernatants at 12, 24 and 36 h posttransfection, centrifuge at $1,000 \times g$ 20 min to remove cell debris, and filter through a sterile 0.45 µm filter; store vectors at –80°C in aliquots before use. Further concentration of vectors by centrifugation may be performed as described (28). Virus should be stored in aliquots at -80°C.

5. Titrate LVs by infecting TE671 at serial dilutions of 10^{-4}, 10^{-5} and 10^{-6} as described (29). Briefly, TE671 cells are plated in a 12-well plate and infected with the diluted LVs in duplicates in the presence of polybrene (10 µg/ml) in a medium volume of 300 µl per well. The plate is gently tilted every hour for 5 h. After two days, the cells are stained for the lentiviral reporter gene expression. Alternatively, the genomic DNA of the infected cells is harvested and the integrated lentiviral proviral DNA is quantitatively assessed by polymerase chain reaction.

3.2. Reporter PLAP Assay

1. Transduced cells are washed with PBS 2–3 times.

2. Fix cells at RT with the formaldehyde fixative for 5 min.

3. Wash cells with PBS 2–3 times, and add 0.5 ml PBS to each well, heat the plate in a 65°C incubator (or in a water bath with the plate tape sealed) for 30 min to inactivate endogenous alkaline phosphatase.

4. Prepare PLAP staining solution using the Alkaline Phosphatase Substrate Kit. Briefly, to a 5 ml 100 mM Tris-HCl (pH 9.5), add a drop or less of Levamisole and mix well; add 2 drops of Reagent 1 and mix well; add 2 drop of Reagent 2 and mix well; add 2 drops of Reagent 3 and mix well (see Note 2).

5. Remove PBS from the wells, add 3 ml of PLAP staining solution and incubate at RT for 0.5–3 h away from light. If the reaction is weak, leave the reaction overnight at room temperature. If the reaction is strong, move the plate to 4°C overnight.

3.3. Preparation of DCs From Peripheral Blood

1. Place fresh heparinized blood into 15- or 50-ml conical centrifuge tubes and dilute with equal volume of PBS; leukocytes from leukapheresis should be diluted with four volumes of PBS.

2. The blood/PBS mixture is gently layered over 1/3 to 1/2 volume of Ficoll–Hypaque solution and centrifuged at 600×g, 18 to 20°C for 25 min.

3. Remove the upper layer that contains plasma and platelets and transfer the peripheral blood mononuclear cell (PBMC) layer to a new centrifuge tube; wash cells three times with two to three volumes of PBS, and centrifuge at 450×g, RT for 10 min.

4. Resuspend cells in AIM-V and determine cell number and viability by trypan blue exclusion (see Note 3).

5. Plate PBMCs at 10^7 cells/2 ml/well in 6-well tissue culture plates and let adhere at 37°C for 2 h (see Note 4).

6. The nonadherent cells are gently removed and cryopreserved as source of immune cells.

7. The adherent cells are cultured for 5 days in AIM-V supplemented with GM-CSF (50 ng/ml) and IL-4 (25 ng/ml). The cells are fed with the same medium with growth factors every other day.

8. On day 5, immature dendritic cells are collected and plated in 24-well low attachment plates in AIM-V supplemented with LPS 1 mg/ml, TNFα (50 ng/ml) and IFN-γ (50 ng/ml) for maturation (see Note 5).

9. Mature DCs are harvested 24–48 h later, and phenotype of mature DCs is analyzed by flow cytometry. Briefly, the cells are stained with anti-CD11c antibody to determine the total DC yield, with anti-CD80, anti-CD86 and anti-CD40 antibodies to monitor for the upregulation of T cell costimulatory molecules, and anti-CD83 antibody to assess the maturation status.

3.4. Lentiviral Transduction of DCs

1. Plate immature DCs into 24-well low attachment plates at 5×10^5 per well with 200 µl of AIM-V supplemented with GM-CSF (50 ng/ml), IL-4 (25 ng/ml) and polybrene (5 mg/ml).

2. Add LVs at multiplicity of infection (MOI) of 20–40; one MOI equals one infectious unit per cell (see in vitro).

3. Incubate cells at 37°C for 2 h, with gentle tilting every 30 min.

4. Add 1 ml of AIM-V supplemented with GM-CSF (50 ng/ml) and IL-4 (25 ng/ml) and continue incubation for an additional 12 h.

5. Induce DC maturation by adding LPS (1 µg/ml), TNFα (50 ng/ml) and IFN-γ (50 ng/ml) for 24–48 h; mature DCs can be harvested and pulsed with peptides at 37°C, 5% CO_2 for 2 h in a 24-well low attachment plate.

3.5. Activation of Antigen-Specific T Cells by LV-Modified DCs

1. Thaw autologous nonadherent PBMCs in AIM-V medium and count cell number and determine viability by trypan blue exclusion.

2. LV-modified and/or peptide-loaded mature DCs in AIM-V are irradiated with 2,000 rad. (20 Gy), and washed twice with AIM-V (see Note 7).

3. Mix nonadherent PBMCs and irradiated DCs at a ratio of 20 to 1 and centrifuge at $320 \times g$ for 7 min. Resuspend the mixed cells in AIM-V supplemented with 5% human AB serum (see Note 8).

4. Plate cells at $6 \times 10^6/2$ ml/well in AIM-V in a 24-well plate (see Note 9) and on day 3, replace half of the medium with fresh medium containing IL-2 (12.5 U/ml), IL-7 (10 ng/ml) and IL-15 (20 ng/ml), and the culture is fed every other day with fresh cytokine-containing medium.

5. Split cells into two wells as necessary (see Note 10).

6. After incubation for 10–20 days, the cells can be analyzed for antigen specific effector functions.

3.6. MHC-Peptide Multimer Binding Analyses of Antigen-Specific CD8 Effector T Cells

1. Wash the cultured T cells twice and resuspend in FACS buffer at a concentration of 5×10^5 cells/ml.

2. Add 200 µl of sample into a flow tube for each multimer assay.

3. To each tube, add appropriately titrated multimer; for MHC class I tetramer, incubate at room temperature for 30 min; for MHC class I pentamer, incubate at room temperature for 15 min; for MHC class II tetramer, incubate at 37°C in 5% CO_2 for 2 h (see Note 11).

4. Add 3 ml of cold FACS buffer, centrifuge at $400 \times g$ for 5 min, discard supernatant and resuspend cells in 100 µl of cold FACS buffer.

5. Add labeled anti-human CD4, CD8 and CD3 Abs and incubate at 4°C for 30 min; dilute with 3 ml of FACS buffer and centrifuge at $400 \times g$ for 5 min (see Note 12).

6. Wash cells twice and resuspend in 500 µl of cold fixation FACS buffer for analysis.

4. Notes

1. All of the plasmids are freely distributed by NIH AIDS Reference and Reagent Program (https://www.aidsreagent.org/Index.cfm) donated by the author; the plasmid DNA quality is critical to high efficiency transfection and high titer vector production.

2. The human placental alkaline phosphatase gene, PLAP, is heat stable and is resistant to some chemical inhibitors that are active on other endogenous alkaline phosphatases. Background activities can be minimized with heat. Histochemical staining for PLAP yields precipitated and highly colored products.

3. The washing step usually removes platelets. With excess platelet contamination, layer PBMCs ($1–2 \times 10^7$ cells/ml) over 3 ml of FBS, centrifuge at $200 \times g$ for 15 min and discard the supernatant containing platelets.

4. If PBMCs are collected from G-CSF mobilized peripheral blood, due to high monocyte content, less cells (about $6–8 \times 10^6$) should be plated into 6-well plates.

5. Highly adherent immature DCs can be treated with 5 mM EDTA in AIM-V at 37°C for 20 min to increase recovery.

6. The MOI is determined based on the titer of the LVs used. For example, the infection of 10^6 cells with 1 µl of 10^9 infectious unit/ml of LVs (equals 10^6 infectious units) is an MOI of 1.

7. Irradiation does not affect DC's antigen presentation functions. The irradiation of DCs is for the purpose of reducing prolonged activation of the cocultured T cells, which may induce activation-induced T cell death (AICD). In addition, lentiviral modification of DCs may confer genetic instability of the DCs. This concern is diminished when the DCs are irradiated.

8. When enriched T cells (for examples, CD4+, CD8+ or CD3+ T cells) are used for coculture with DCs, the ratio of T cells to DCs should be adjusted to 10:1 (30).

9. During the initial 3–4 days of coculture, do not disturb the cells to ensure the quality of synapse formation of DCs and T cells.

10. The growth of T cells is dependent on appropriate cell density. If the cell number is low, cells should be plated into U-bottom 96-well plate at 5×10^5 cells per well.

11. It is important to titrate each multimer before use.

12. Some human CD8 mAb clones may interfere with the binding of multimer to T cell receptor (31).

Acknowledgments

I thank W. Chou, B. Wang, and S. Han for technical assistance. The presented work was supported by NIH grants and funds from Yong-Ling Foundation, Taiwan.

Disclosures

L.-J. Chang has declared a financial interest as consultant to a company and holds patents related to the work that is described in the present study.

References

1. Guermonprez, P., Valladeau, J,, Zitvogel, L., Thery, C., and Amigorena, S. (2002) Antigen presentation and T cell stimulation by dendritic cells. *Annu Rev Immunol* **20**, 621–67.

2. Ridgway, D. (2003) The first 1000 dendritic cell vaccinees. *Cancer Invest* **21**, 873–86.

3. Girolomoni, G., and Ricciardi-Castagnoli, P. (1997) Dendritic cells hold promise for immunotherapy. *Imm. Today* **18**, 102–4.

4. van Schooten, W.C., Strang, G., and Palathumpat, V. (1997) Biological properties of dendritic cells: implications to their use in the treatment of cancer. *Mol Med Today* **3**, 254–60.

5. Toes, R.E., van der Voort, E.I., Schoenberger, S.P., Drijfhout, J.W., van Bloois, L., Storm, G., Kast, W.M., et al. (1998) Enhancement of tumor outgrowth through CTL tolerization after peptide vaccination is avoided by peptide presentation on dendritic cells. *J Immunol* **160**, 4449–56.

6. Lanzavecchia, A., and Sallusto, F. (2001) Regulation of T cell immunity by dendritic cells. *Cell* **106**, 263–6.

7. Miller, P.W., Sharma, S., Stolina, M., Chen, K., Zhu, L., Paul, R.W., and Dubinett, S.M. (1998) Dendritic cells augment granulocyte-macrophage colony-stimulating factor (GM-CSF)/herpes simplex virus thymidine kinase-mediated gene therapy of lung cancer. *Cancer Gene Ther* **5**, 380–9.

8. Pulendran, B., Banchereau, J., Maraskovsky, E., and Maliszewski, C. (2001) Modulating the immune response with dendritic cells and their growth factors. *Trends Immunol* **22**, 41–7.

9. Wan, Y., Bramson, J., Pilon, A., Zhu, Q., and Gauldie, J. (2000) Genetically modified dendritic cells prime autoreactive T cells through a pathway independent of CD40L and interleukin 12: implications for cancer vaccines. *Cancer Res* **60**, 3247–53.

10. Gabrilovich, D. (2004) Mechanisms and functional significance of tumour-induced dendritic-cell defects. *Nat Rev Immunol* **4**, 941-52.

11. Chen, X., He, J., and Chang, L.J. (2004) Alteration of T cell immunity by lentiviral transduction of human monocyte-derived dendritic cells. *Retrovirology* **1**, 37.

12. Breckpot, K., Aerts, J.L., and Thielemans, K. (2007) Lentiviral vectors for cancer immunotherapy: transforming infectious particles into therapeutics. *Gene Ther* **14**, 847–62.

13. Ludewig, B., Ehl, S., Karrer, U., Odermatt, B., Hengartner, H., and Zinkernagel, R.M.. (1998) Dendritic cells efficiently induce protective antiviral immunity. *J Virol* **72**, 3812-8.

14. Kirk, C.J., and Mule, J.J. (2000) Gene-modified dendritic cells for use in tumor vaccines. *Hum Gene Ther* **11**, 797–806.

15. Jenne, L., Schuler, G., and Steinkasserer, A. (2001) Viral vectors for dendritic cell-based immunotherapy. *Trends Immunol* **22**, 102–7.

16. Murphy, A., Westwood, J.A., Teng, M.W., Moeller, M., Darcy, P.K., and Kershaw, M.H. (2005) Gene modification strategies to induce tumor immunity. *Immunity* **22**, 403–14.

17. Lipscomb, M.F., and Masten, B.J. (2002) Dendritic cells: immune regulators in health and disease. *Physiol Rev* **82**, 97–130.

18. Mosmann, T.R., and Coffman, R.L. (1989) TH1 and TH2 cells: different patterns of lymphokine secretion lead to different functional properties. *Ann Rev Immunol* **7**, 145–73.

19. Kalinski, P., Hilkens, C.M., Wierenga, E.A., and Kapsenberg, M.L.. (1999) T-cell priming by type-1 and type-2 polarized dendritic cells: the concept of a third signal. *Immunol Today* **20**, 561–7.

20. Condon, C., Watkins, S.C., Celluzzi, C.M., Thompson, K., and Falo, L.D.J. (1996) DNA-based immunization by in vivo transfection of dendritic cells. *Nature Med* **2**, 1122–8.

21. Liu, M. Transfected human dendritic cells as cancer vaccines. (1998) *Nature Biotechnol* **16**, 335–6.

22. Wang, B., He. J,. Liu. C., and Chang, L.J. (2006) An effective cancer vaccine modality: Lentiviral modification of dendritic cells expressing multiple cancer-specific antigens. *Vaccine* **24**, 3477–89.

23. Chambers, C.A. (2001) The expanding world of co-stimulation: the two-signal model revisited. *Trends Immunol* **22**, 217–23.

24. Chirathaworn, C., Kohlmeier, J.E., Tibbetts, S.A., Rumsey, L.M., Chan, M.A., and Benedict, S.H. (2002) Stimulation through intercellular adhesion molecule-1 provides a second signal for T cell activation. *J Immunol* **168**, 5530–37.

25. Salomon, B., and Bluestone. J.A. (1998) LFA-1 interaction with ICAM-1 and ICAM-2 regulates Th2 cytokine production. *J Immunol* **161**, 5138–5142.

26. Sabatte, J., Maggini, J., Nahmod, K., Amaral, M.M., Martinez, D., Salamone, G., Ceballos, A., et al. (2007) Interplay of pathogens, cytokines and other stress signals in the regulation of dendritic cell function. *Cytokine Growth Factor Rev* **18**, 5–17.

27. Han, S., Wang, B., Cotter, M.J., Yang, L.J., Zucali, J., Moreb, J.S., and Chang, L.J. (2008) Overcoming immune tolerance against multiple myeloma with lentiviral calnexin-engineered dendritic dells. *Mol Ther* **16**, 269–79.

28. Chang L.J., and Zaiss A.K. (2001) Methods for the preparation and use of lentivirus vectors. In: Morgan J, (ed). *Gene Therapy Protocols*. **vol, 2** 2nd edn: Humana Press, Inc., 303–18.

29. Chang, L.J., and Zaiss, A.K. (2001) Self inactivating lentiviral vectors in combination with a sensitive Cre/loxP reporter system. In: Walker J, ed. *Methods in Molecular Medicine*. Humana Press Inc., 367–82.

30. Hopken, U.E., Lehmann, I., Droese, J., Lipp, M., Schuler, T., and Rehm, A. (2005) The ratio between dendritic cells and T cells determines the outcome of their encounter: proliferation versus deletion. *Eur J Immunol* **35**, 2851–63.

31. Denkberg, G., Cohen, C.J., and Reiter, Y. (2001) Critical role for CD8 in binding of MHC tetramers to TCR: CD8 antibodies block specific binding of human tumor-specific MHC-peptide tetramers to TCR. *J Immunol* **167**, 270–76.

32. Iwakuma, T., Cui, Y., and Chang, L.J. (1999) Self-inactivating lentiviral vectors with U3 and U5 modifications. *Virology* **261**, 120–32.

Part IV

The Short RNA Technologies Applied by Lentiviral Vectors

Chapter 12

Lentiviral Vector-Mediated Expression of pre-miRNAs and AntagomiRs

Michaela Scherr, Letizia Venturini, and Matthias Eder

Abstract

Micro (mi)RNAs are highly conserved small regulatory RNAs, which regulate gene expression by hybridization to specific binding sites in the 3′untranslated region (UTR) of many mRNAs. Upon miRNA-guided recruitment of a multiprotein complex, target mRNAs are either degraded or their translation is blocked depending on the complementarity between the miRNAs and their binding sites in target mRNAs. Individual miRNAs have been shown to regulate the expression of hundreds of genes with corresponding miRNA binding sites in the 3′UTR in a dose-dependent manner. Although miRNA-target genes may be predicted by bioinformatic tools, each potential target needs to be confirmed experimentally. We describe here the expression of individual miRNAs or miRNA-specific antagomiRs by lentiviral gene transfer to induce stable gain- and loss-of-function phenotypes. These techniques provide some tools to analyze miRNA function in cell culture or animal models.

Key words: AntagomiR, miRNA, RNAi, miRNA silencing, Gene transfer, Lentivirus

1. Introduction

miRNAs are endogenous regulatory RNAs of about 22 nt in length. They constitute a class of regulatory genes that encompasses about 1% of all genes in vertebrates. miRNAs regulate gene expression post-transcriptionally usually by incomplete hybridization to often multiple miRNA-binding sites in the 3′UTR of target mRNAs. The miRNA-mediated guidance of so-called RNA-induced silencing complexes (RISC) to target mRNAs induces either translational repression or mRNA degradation (Fig. 1) (1). Recently, the mild down-regulation of hundreds of proteins by individual miRNAs has been demonstrated, and the functional role of the 6–8 nt long miRNA sequence hybridizing

Maurizio Federico (ed.), *Lentivirus Gene Engineering Protocols*, Methods in Molecular Biology, vol. 614,
DOI 10.1007/978-1-60761-533-0_12, © Humana Press, a part of Springer Science+Business Media, LLC 2010

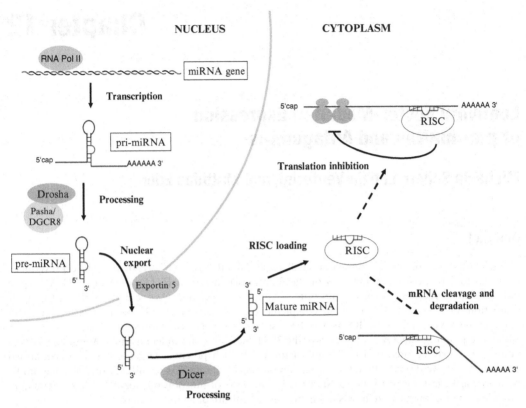

Fig. 1. Schematic representation of miRNA biogenesis and mechanism of function. miRNA-encoding genes are transcribed by RNA Polymerase II as large pri-miRNA transcripts with 5′cap and 3′ poly-A tail. In the nucleus, they are processed to hairpin precursors (pre-miRNAs) of about 70 nts by a protein–processor complex containing the RNase III Drosha and the double-stranded DNA-binding protein Pasha/DGCR8. Following transport to the cytoplasm by Exportin 5, the pre-miRNA is cleaved by Dicer, another RNase III enzyme, into 19–22-nt length mature double-stranded miRNA. After loading the antisense strand into the RNA-induced silencing complex (RISC), it guides RISC to sequence-specific recognition of the target mRNA leading to translational inhibition or mRNA degradation, depending on partial or perfect homology, respectively

to mRNA targets, the so-called seed sequence, has evolutionarily and functionally been defined (2–4). Interestingly, miRNAs are involved in development and immunity, and altered miRNA expression has also been described in a variety of human malignancies (5). However, the functional relevance of aberrant miRNA expression for tumorigenesis has still to be defined in most cases.

miRNAs are processed in a regulated multistep process (6) schematically shown in Fig. 1. Long primary transcripts (pri-miR-NAs) are transcribed from Pol II promoters either from individual – sometimes polycistronic – miRNA genes or from introns of protein-coding genes. Pri-miRNAs are processed in the nucleus by Drosha, an RNaseIII-type nuclease, in complex with DGCR8 into ~70 nt long precursor miRNAs (pre-miRNAs). Upon cytoplasmic export by Exportin-5, pre-miRNAs are further processed by Dicer, another RNaseIII-type enzyme, into mature double-stranded

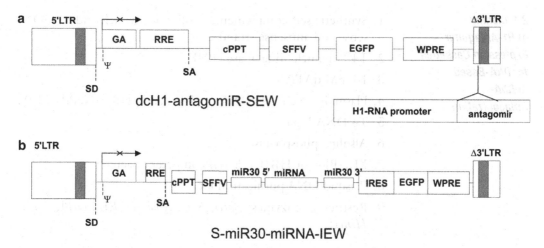

Fig. 2. Lentivirus-mediated antagomiR- and miRNA-expression. (**a**) Schematic representation of the lentiviral trans-gene-plasmid harboring an antagomir expression cassette with a human H1-RNA promoter inserted into the U3 region of the Δ3′LTR. This location results in duplication during reverse transcription indicated as "double-copy" (dc) vector. S:SFFV-LTR, E:EGFP, W:WPRE. (**b**) Schematic representation of the lentiviral transgene-plasmid containing a miRNA-expression cassette. Specific miRNA-sequences are embedded within sequences derived from miR-30. S:SFFV-LTR, I: IRES, E:EGFP, W:WPRE

miRNAs. Similar to siRNAs, one miRNA strand is incorporated into a multiprotein complex RISC (RNA-induced silencing complex), which is recruited by the miRNA to target mRNAs. The complimentarity between mRNA and miRNA binding site in the 3′UTR (and may be other factors still to be defined) determines the quality and kinetics of miRNA/RISC – mRNA interaction, leading either to translational repression and/or mRNA deadenylation and decay.

We describe here lentiviral gene transfer strategies to induce stable gain- and loss-of-function phenotypes for individual miRNAs. Upon transduction, lentivirally encoded expression cassettes either for miRNA expression or for transcription of RNAs complementary to specific miRNAs (so-called antagomiRs) are integrated into the host cell genome resulting in long-term modulation of endogenous miRNA levels. As shown in Fig. 2, antagomiRs can be expressed from Pol III promoters, such as H1, as short RNAs, whereas for miRNA overexpression, specific miRNAs are embedded into heterologous sequences derived from miR-30, for example, under control of Pol II promoters (7).

2. Materials

The protocols outlined below describe the use of lentivirally transcribed antagomiRs (miRNA inhibitor) and pre-miRNAs (miRNA precursor), in a mammalian cell culture system.

2.1. Cloning of H1-Antagomir Expression Cassettes for DNA-Based miRNA-Antagonization

1. Synthetic self-complementary oligodeoxynucleotides (ODN; sense and antisense strand).

2. T4 Polynucleotidekinase (PNK).

3. 10 mM dATP.

4. Plasmids: pSUPER (DNA$_{engine}$ Inc.) and pHR-SINcPPT-SEW.

5. T4 DNA Ligase.

6. Alkaline phosphatase.

7. XL1-Blue or HB101 *E. coli* competent bacteria.

8. Plasmid DNA purification kit.

9. Restriction enzymes: *BglII, SalI, XhoI, EcoRI, SnaBI, SmaI, HincII*.

2.2. Cloning of Pre-miRNA Expression Cassettes for DNA-Based miRNA-OverExpression

1. Synthetic oligodeoxynucleotides (ODN; sense and antisense strand).

2. T4 Polynucleotidekinase (PNK).

3. 10 mM dATP.

4. Plasmids: pSUPER-miR30 and pHR-SINcPPT-SIEW.

5. T4 DNA Ligase.

6. Alkaline phosphatase.

7. Klenow DNA polymerase I.

8. XL1-Blue or HB101 E. coli competent bacteria.

9. Plasmid DNA purification kit.

10. Restriction enzymes: *EcoRI, BamHI, BglII, XhoI*.

2.3. Generation of Lentiviral Vectors

1. Cell lines: human embryonal kidney 293 cell line (DSMZ No. ACC 305, 293 T cells can be used as well) for lentiviral vector production. The human chronic myeloid leukaemia K562 cell line (DSMZ No. ACC 10) for lentiviral vector titration and functional studies.

2. Media: DMEM (293 cells) and RPMI (K562 cells) supplemented with 10% FCS.

3. Transgene plasmid carrying the pre-miRNA or the antagomiR expression cassette.

4. HIV-1 packaging vector: pCMVΔR8.9.1 (8).

5. Envelope expressing vector: pMD.G (containing the glycoprotein G of the vesicular stomatitis virus-VSV.G).

6. Transfection buffer: (a) stock solution of Na$_2$HPO$_4$ dibasic (5.25 g in 500 ml water). (b) 2 × HBS: 8 g NaCl, 6.5 g HEPES (sodium salt), 10 ml Na$_2$HPO$_4$ from the stock solution. Bring to pH 6.95–7.0 using NaOH or HCl and bring volume up to 500 ml.

7. 2M CaCl$_2$.

8. 100× protamine sulfate dissolved in water (400 µg/ml).

9. 0.01% (w/v) Poly-L-lysine dissolved in water.

10. Phosphate-buffered saline (PBS; pH 7.4, Ca^{2+}- and Mg^{2+}-free).

2.4. RNA Isolation and Detection of miRNAs by miRNA-Specific Quantitative RT-PCR (miR-qRT-PCR)

1. Trizol reagent (Invitrogen Ltd. Paisley, UK).

2. miRNA-specific looped RT-primers and Taqman probes (Applied Biosystems, Darmstadt, Germany).

2.5. Immunoblotting

1. Lysis buffer: 250 mM Tris–HCl, pH7.5; 0.5% TritonX-100; 10 mM NaF; 10 mM Na$_3$VO$_4$; 1 mM PMSF; 2 mM EDTA; 2 mM AEBSF; 2 µM Aprotinin; 100 µM Bestatin; 30 µM E-64; 40 µM Leupeptin; 20 µM Pepstatin A.

2. SDS-PAGE.

3. Hybond enhanced chemiluminescence (ECL) nitrocellulose membrane (Amersham Bioscience, Uppsala, Sweden).

4. Monoclonal mouse anti-E2F-1 (KH20+KH95, Upstate, Hamburg, Germany), monoclonal mouse anti-α-tubulin (DM1A, Calbiochem, Schwalbach am Ts, Germany).

5. ECL Western blotting detection reagents (Amersham Biosciences, Uppsala, Sweden).

3. Methods

miRNA inhibitor molecules (antagomiRs) are single-stranded RNA molecules that specifically hybridize to and inhibit endogenous miRNAs. miRNA precursor molecules (pre-miRNA) and antago-miRs can be generated either by chemical synthesis or transcribed from an expression plasmid harboring a pre-miRNA and antago-miR cassette, respectively. Overexpression of pre-miRNAs enables analysis of miRNA biological effects (gain-of-function experiments) and the use of antagomiRs enables functional analysis by specific inhibition of miRNA activity (loss-of-function experiments).

3.1. Cloning of H1-Antagomir Expression Cassettes for DNA-Based miRNA Antagonization

Single-stranded ODNs (~40 nucleotides in length) encoding a 21–25 nt antagomiR-ODN (antisense strand to the corresponding miRNA *italic*), a termination signal consisting of six thymidines, and overhang sequences from a 5′ *Bgl*II- and a 3′ *Sal*I-restriction site, can be designed as follows: sense-antago-miR: 5′-GATCCCC (N)$_{21-25}$ TTTTTTGGAAG-3′, antisense-antagomiR: 5′-TCGACTTCCAAAAAA (N′)$_{21-25}$ GGG-3′. They are phosphorylated and hybridized with the corresponding complementary single-stranded ODN. The resulting dsDNA is inserted into the *Bgl*II–*Sal*I site of the dephosphorylated plasmid pSUPER (suppression of endogenous RNA (9) harboring the H1-RNA promoter to generate pSUPER-antagomiR. *E. coli* HB101 or XL1-Blue competent cells are transformed with the resultant plasmids, followed by plasmid DNA purification. Both the correct sequence and insertion are confirmed by DNA sequencing for each plasmid.

3.1.1. Phosphorylation and Annealing of ODN

1. Dissolve ODN in water to obtain a final concentration of 1 mM.
2. Take 1 µl from each ODN (sense- and anti-sense-antagomiR).
3. Add 1 µl 10×T4 PNK buffer.
4. Add 1 µl 10 mM ATP.
5. Add 1 µl T4 PNK (10 U/µl).
6. Add 5 µl H$_2$0.
7. Incubate 30–45 min at 37°C.
8. Denature at 95°C for 3 min and cool slowly down to room temperature to anneal ODN.

3.1.2. Ligation of the Duplex in pSUPER Plasmid

1. Take 3 µl of the duplex (annealing reaction).
2. Add 1 µl 10×T4 ligase buffer.
3. Add 1 µl dephosphorylated plasmid pSUPER (digested with *Bgl*II and *Sal*I).
4. Add 1 µl T4 DNA Ligase (400 U/µl).
5. Add 4 µl H$_2$O.
6. Incubate 1 h at room temperature.
7. Transform *E. coli* HB101 or XL1-Blue competent cells, and recover plasmid DNA.
8. Analyze the insert by *Eco*RI/*Xho*I digestion with positive clones identified by inserts of ~320 bp for the plasmid pSUPER-antagomiR (see Note 1).

3.2. Cloning of Pre-miRNA-Expression Cassettes for DNA-Based OverExpression

1. Phosphorylate and hybridize the single-stranded ODN (~70–120 nucleotides in length) encoding the sequence of the corresponding pre-miRNAs and overhang sequences from a 5′ *Eco*RI- and a 3′ *Bam*HI- restriction site with the corresponding complementary single-stranded ODN.
2. Insert the resulting dsDNA into the *Eco*RI–*Bam*HI site of the dephosphorylated plasmid SUPER-miR-30 containing the miR-30 cassette to generate pSUPER-miR-30-miRNA (this vector is used to clone the heterologous miR-30 cassette into suitable viral vectors only) (9).

3. Transform *E. coli* XL1-Blue competent cells with the resultant plasmid, and recover the plasmid DNA.

4. Confirm the correct sequence and insertion by *Eco*RI/*Bam*HI digestion, generating a ~200 base pairs insert, and DNA sequencing.

3.3. Stable OverExpression and Silencing of miRNAs Through Lentiviral Vectors

Upon transfection by electroporation or lipofection, gene silencing is only transient depending on cell cycle kinetics of the target cells, the half-life of the target protein, and the turnover of the target mRNA. Suitable technologies to induce stable or long-term expression employ different classes of viral vectors to transduce a variety of cell types with high efficiency. Retro/lentiviruses can infect a wide variety of cell types and integrate as proviruses into the host cell genome. In addition, expression of RNAi triggers upon retro/lentiviral gene transfer can be used for studies in primary cells, e.g., hematopoietic cells, which are usually difficult to transfect. Since lentiviral vectors are able to integrate into the host genome of nondividing cells, such as stem cells or terminally differentiated cells like neurons, the range of target cells is expanded as compared to retroviral delivery of RNAi triggers independent of viral pseudotyping. As illustrated in Fig. 2a, the H1-antagomiR expression cassettes can be inserted into the lentiviral transgene plasmid dcH1-antagomiR-SEW (dc:double-copy, S:SFFV promoter, E:EGFP, W:WPRE a posttranscriptional regulatory element of woodchuck hepatitis virus) within the U3 region of the 3'-LTR. The location in the U3 region of the 3'-LTR leads to a duplication of the H1-cassette during reverse transcription. Since endogenously transcribed miRNAs are usually transcribed from polymerase II (pol II) promoters, the miR-30-based miRNA expression cassette should be inserted into the bi-cistronic lentiviral transgene plasmid SIEW (I=internal ribosome entry site) 3' of the internal pol II SFFV promoter as illustrated in Fig. 2b. (see Note 2).

To construct the double-copy variant dcH1-antagomiR-SEW, the plasmid dcH1-SEW should be digested with *Sna*BI followed by dephosphorylation (10). H1-antagomiR cassettes are released from pSUPER-antagomiR by digestion with *Sma*I and *Hin*cII.

3.3.1. Cloning of Lentiviral Vectors Expressing AntagomiR or miR-30-Based-miRNA

1. Take 1 μl (~400 ng) H1-antagomiR or miR30-miRNA-fragment.

2. Add 1 μl 10×T4 ligase buffer.

3. Add 1 μl (100 ng) digested and dephosphorylated plasmid dcH1-SEW (for antagomiR expression) or SIEW (for miR-30-miRNA expression).

4. Add 1 μl T4 DNA Ligase (400 U/μl).

5. Add 6 μl H_2O.

6. Incubate 3–4 h at room temperature or overnight at 16°C.

7. Transform *E. coli* HB101 or XL1-Blue competent cells, and isolate the respective plasmids (see Note 3).

8. Inserts can be analyzed by *Pst* I digestion with positive clones identified by inserts of ~7,800, 1,180, and 890 bp for the plasmid dcH1-antagomiR, and by digestion with *XhoI/NcoI* resulting in fragments of ~5,280, 3,350, 1,320, and 610 bp for S-miR30-miRNA-IEW, respectively.

3.4. Generation of Lentiviral Supernatants

3.4.1. Generation of Lentiviral Vector Particles

1. Three days before transfection, plate 6×10^6 293 cells onto a poly-L-lysine (0.01% solution) coated T175 flask in 25 ml DMEM.

2. Transfect the subconfluent 293 cells with 60 μg of the lentiviral transgene plasmid, 45 μg packaging- and 30 μg envelope-plasmid DNA using the calcium phosphate method (11).

3. After 16 h, wash the cells with 1 × PBS (pH 7.4, Ca^{2+}- and Mg^{2+}-free) and incubate for additional 8 h in fresh medium.

4. Collect the culture supernatant containing the lentiviral particles 24–72 h after transfection.

5. Filter through a 0.45 μm pore size filter to remove cell debris, and concentrate the viral particles by low-speed centrifugation at 10.000×*g*, 10°C for 16 h.

6. For titer determination transduce 10^5/ml 293 cells in a 24-well plate with serial dilutions (ranging from 10^{-2} to 10^{-9}) of concentrated viral particles. Three days later, the transduced 293 cell populations are analyzed by FACS. Titers are determined using the highest dilution for which cells containing GFP-fluorescence above background are visible in FACS analysis. Store the virus in aliquots at –80°C. (see Note 4).

3.5. RNA Isolation and Detection of miRNAs by miRNA-Specific Quantitative RT-PCR (miR-qRT-PCR)

The expression and inhibition of specific miRNAs can be measured by miR-qRT-PCR using miRNA-specific looped RT-primers and TaqMan probes as recommended by the manufacturer. Normalization can be performed using the 2-ΔCT method relative to U6snRNA or an endogenous miRNA whose expression is constant in the cells under investigation (see Note 5).

3.5.1. Transduction of K562 Cells

0.5×10^6 K562 cells are transduced in a 24-well plate by adding 5 μl of lentivirus (corresponding to approx.10^7 of lentiviral vector particles) (see Note 6) and 5 μl protamine sulfate (final concentration 4 μg/ml) in a total volume of 500 μl (multiplicity of infection; MOI ~10) followed by spinoculation at 2,000×*g* at 32°C for

90 min and over-night incubation at 37°C. After transduction, the supernatants are removed, and cells are maintained in RPMI/10% FCS. Transduction efficiency is monitored by the FACS analysis of GFP expression on day 4 (and later on).

3.5.2. Total Cellular RNA Extraction From Transduced K562 Cells and Analysis of miRNA Expression and Inhibition by miR-qRT-PCR

1. Isolate total cellular RNA from the transduced K562 cells, for example using the Trizol reagent.

2. Use 10 ng of total cellular RNA, 50 nM stem-loop RT-primer, RT-buffer, 0.25 mM dNTP (each), 3 units/ml MultiScribe reverse transcriptase, and 0.25 units/ml RNAse inhibitor in a 15 μl reaction for 30 min at 16°C, 30 min at 42°C, and 5 min at 85°C.

3. For real-time PCR, add 1.5 μl of cDNA, 0.2 mM Taqman probe, 1.5 mM primers (reverse and forward), and TaqMan Universal PCR Master Mix to a total of 20 μl of reaction volume for 10 min at 95°C and 40 cycles of 15 s at 95°C and 1 min at 60°C. Amplify on an ABI 7500 cycler.

3.6. Monitoring of miRNA OverExpression or miRNA-Inhibition by Immunoblotting of a Known Specific Target Gene Product

The impact of lentivirally encoded antagomiRs or pre-miRNAs on miRNA-regulated protein expression can be studied by immunoblotting. For example, E2F-1 is a confirmed cellular target of miR-20a (12). To analyze the impact of antagomiR 20a on miRNA-regulated protein expression, K562 cells can be transduced with either anti-miR-20a antagomiRs or miR-20a. In a typical experiment, expression of antagomiRs and miR-20a demonstrated a –4.2- and a +2.2-fold induction of miR-20a levels detected by miR-qRT-PCR, respectively. Whole cell lysates of transduced K562 cells were analyzed for E2F-1 expression by immunoblotting. As shown in Fig. 3, expression of miR-20a (lane 6) reduces, whereas that of anti-miR-20a antagomiRs (lane 3) increases E2F-1 levels.

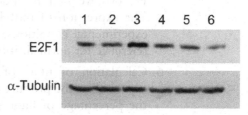

Fig. 3. Effects of lentivirally encoded miR-20a and anti-miR-20a antagomir expression on E2F-1 expression. K562 cells transduced with control dcH1-ctrl-antagomir-SEW (*lane 2*), dcH1-antagomir-20a-SEW (*lane 3*), S-mir30-ctrl-IEW (*lane 5*), and S-mir30-20a-IEW (*lane 6*), respectively were lysed 4 days after transduction. *Lanes 1* and *4* show untransduced K562 cells. The immunoblots were probed with an anti-E2F-1 antibody, and reprobed with an anti-α-tubulin antibody as loading control

3.6.1. Immunoblotting

1. Prepare total cellular lysates from transduced K562 cells with lysis buffer.

2. Subject 20 μg of total cellular lysates to electrophoresis in a polyacrylamide gel containing sodium dodecyl sulfate.

3. Transfer to Hybond enhanced chemiluminescence (ECL) nitrocellulose membrane.

4. Probe the blots with monoclonal-mouse anti-E2F1 and monoclonal mouse anti-α-tubulin antibodies.

5. Visualization is performed using the ECL Western blotting reagents according to the manufacturer's recommendations.

4. Notes

1. It is recommended to include a "false" antagomiR with a scrambled sequence and a vector alone control in testing antagomiR constructs.

2. It is important to consider the type of pol II or pol III promoter to be used depending on the study design. Both pol II promoters, such as CMV, EF-1α, and pol III promoters like U6 and H1-RNA, have been successfully employed for si/shRNA expression and knock-down experiments.

3. The use of recombination-deficient bacteria (such as HB101 or XL1-Blue) is crucial for successful cloning of lentiviral plasmids.

4. Production of high-titer lentiviral supernatants may require some training and experience. Make sure that the cells used for the lentiviral vector production are free of contamination by mycoplasms.

5. For K562 and other human cell lines (like as EM-2, LAMA-84, HL-60), we used the endogenous miR-16 for normalization. The expression of miR-16 was almost constant under the experimental conditions, a prerequisite for a good internal reference for normalization of the results.

6. Calculation of titers of lentiviral supernatants: titers are expressed as the product of the dilution factor multiplied by the percentage of fluorescent cells, then multiplied by the number of cells transduced.

Acknowledgment

Supported in part by grants of the "Deutsche Forschungsgemeinschaft" (SFB 566 and REBIRTH) and H.W. & J. Hector-Stiftung.

References

1. Shyu, A.B., Wilkinson, M.F., and van Hoof, A. (2008) Messenger RNA regulation: to translate or to degrade *EMBO J* **27**, 471–81.

2. Lewis, B.P., Burge, C.B., and Bartel, D.P. (2005) Conserved seed pairing, often flanked by adenosines, indicates that thousands of human genes are microRNA targets *Cell* **120**, **15**–20.

3. Selbach, M., Schwanhäusser, B., Thierfelder, N., Fang, Z., Khanin, R., and Rajewsky, N. (2008) Widespread changes in protein synthesis induced by microRNAs *Nature* **455**, 58–63.

4. Baek, D., Villén, J., Shin, C., Camargo, F.D., Gygi, S.P., and Bartel, D.P. (2008) The impact of microRNAs on protein output *Nature* **455**, 64–71.

5. Lu, J., Getz, G., Miska, E.A., Alvarez-Saavedra, E., Lamb, J., Peck, D., Sweet-Cordero, A., Ebert, B.L., Mak, R.H., Ferrando, A.A., Downing, J.R., Jacks, T., Horvitz H.R., and Golub, T.R. (2005) MicroRNA expression profiles classify human cancers *Nature* **435**, 834–8.

6. Bartel, D.P. (2004) MicroRNAs: genomics, biogenesi000s, mechanism, and function *Cell* **116**, 281–97.

7. Scherr, M., Venturini, L., Battmer, K., Schaller-Schoenitz, M., Schaefer, D., Dallmann, I., Ganser, A., and Eder, M. (2007) Lentivirus-mediated antagomir expression for specific inhibition of miRNA-function *Nuc Acids Res.* **35**, e149.

8. Miyoshi, H., Smith, K.A., Mosier, D.E., Verma, I.M., and Torbett, B.E. (1999) Transduction of human CD34+ cells that mediate long-term entgraftment of NOD/SCID mice by HIV vectors. *Science* **283**, 682–6.

9. Brummelkamp, T.R., Bernards, R., and Agami, R. (2002) A system for stable expression of short interfering RNAs in mammalian cells Science **296**, 550–3.

10. Scherr, M., Battmer, K., Ganser, A., and Eder, M. (2003) Modulation of gene expression by lentiviral-mediated delivery of small interfering RNA *Cell Cycle* **2**, 251–7.

11. Graham, F.L., and van der Eb, A.J. (1973) Transformation of rat cells by DNA of human adenovirus 5 *Virology* **54**, 536–9.

12. O'Donnell, K.A., Wentzel, E.A., Zeller, K.I., Dang, C.V., and Mendell, J.T. (2005) c-Myc-regulated microRNAs modulate E2F1 expression *Nature* **435**, 839–43.

References

Chapter 13

The RNA Silencing Technology Applied by Lentiviral Vectors in Oncology

Hidetoshi Sumimoto and Yutaka Kawakami

Abstract

Since the discovery of RNA interference (RNAi) in *Caenorhabditis elegans* in 1998, this mechanism has been found to be conserved in a wide variety of species, including insects, plants, and mammals. In mammals, small (or short) interfering RNA (siRNA) or short hairpin RNA (shRNA) can be expressed by using several expression vectors including lentiviral vectors. The lentiviral vector has several useful characteristics for RNAi experiments including broad host tropism and stable gene transduction to both dividing and nondividing cells, which permits stable depletion of target genes. This technology can be useful for several applications, including basic cancer research.

Key words: RNAi, shRNA, HIV lentiviral vector, BRAF, SKP-2

1. Introduction

RNAi is a recently discovered posttranscriptional gene silencing (PTGS) mechanism (1), which is conserved among many species including mammals. siRNA, cleaved from a longer double-stranded RNA (dsRNA) by a family of RNase III, Dicer, is an RNA duplex of approximately 21–23 nucleotides (nt) in length, and has a characteristic 2–3 nt overhang in each 3′-end and phosphates at the 5′-ends. The siRNA is incorporated into a protein complex called RNA-induced silencing complex (RISC), and its antisense strand guides the RISC to a target mRNA homologous to the siRNA (2). The slicer activity of AGO2 in the RISC mediates the degradation of the target mRNA, resulting in the PTGS. In mammalian cells, however, dsRNA longer than 30 nt length induces strong interferon responses leading to the global gene silencing and cell toxicity. siRNA less than 30 nt length have

Maurizio Federico (ed.), *Lentivirus Gene Engineering Protocols,* Methods in Molecular Biology, vol. 614,
DOI 10.1007/978-1-60761-533-0_13, © Humana Press, a part of Springer Science+Business Media, LLC 2010

shown to bypass these interferon responses, resulting in the specific gene silencing (3) (see Note 1). siRNA can be expressed as an shRNA form, a fold-back stem loop structure with the sense and antisense RNA, which are linked with a small linker sequence (4). Exportin-5 mediates the export of shRNA from nucleus to cytoplasm, where Dicer degrades the linker sequence, resulting in the siRNA form (5). The following process is the same as the usual RNAi mechanism. shRNA expression cassette can be incorporated into the recombinant HIV lentiviral genome for shRNA expression (6).

We have shown that stable inactivation of several oncogenes with HIV-shRNA vectors results in the amelioration of some malignant phenotypes, such as cell proliferation, invasion, or immune evasion (7–10). In particular, we obtained relevant results through the lentiviral vector-expressed shRNA against BRAF and SKP-2.

BRAF is a mitogen-activated protein kinase kinase kinase (MAPKKK), which is frequently activated by a missense mutation in many human cancers, leading to the constitutively activated MAPK signal. The BRAF mutations were found in 66% of human melanoma cases, and $BRAF^{V600E}$ mutation is the most frequent (80%) of them. The mutated BRAF has more than 10–20-fold higher kinase activity than wild type BRAF , and it is possible to target this disease-specific $BRAF^{V600E}$ allele by using the mutation-specific shRNA vector (7). 526mel, a human melanoma cell line, is heterozygous for $BRAF^{V600E}$. We designed both $mtBRAF^{V600E}$-specific and $wtBRAF^{V600V}$-specific shRNA lentiviral vectors. Although both vectors reduced total BRAF protein levels to the same extent, $mtBRAF^{V600E}$-specific RNAi decreased the cell proliferation significantly when compared with the $wtBRAF^{V600V}$-specific RNAi, which correlated to the degree of ppERK1/2 levels (Fig. 1). $mtBRAF^{V600E}$- shRNA is specific for $BRAF^{V600E}$ since it decreased the BRAF protein level of A375 melanoma cells almost completely, which is homozygous for $BRAF^{V600E}$, but did not decrease the BRAF of 1362mel melanoma cells without $BRAF^{V600E}$. These results indicate the endogenous $BRAF^{V600E}$ is crucial for the cell proliferation of some human melanoma cells (7).

SKP-2, a member of the F-box protein family, is a specific substrate-recognition subunit of an SCF ubiquitin-protein ligase complex and is involved in the CDK inhibitors, $p27^{Kip1}$ and $p21^{Cip1}$, degradation. In many cancers, $p27^{Kip1}$ protein is decreased due to the enhanced degradation, which results in the increase of cell growth. One mechanism is an enhanced ubiquitin-proteasomal pathway, associated with the increase of SKP-2 protein. Inactivation of SKP-2 with RNAi restores the $p27^{Kip1}$ and $p21^{Cip1}$ protein levels, leading to the G1 cell cycle arrest in human cancer cells with SKP-2 overexpression (8, 9). ACC-LC-172, a human small cell lung carcinoma cell line, expresses a large amount of SKP-2

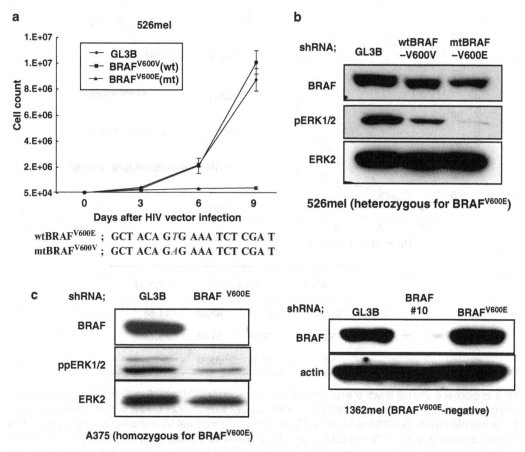

Fig. 1 HIV-mtBRAF^V600E specifically inhibits the mtBRAF^V600E, but not the wt BRAF^V600V. (**a**) 526mel, heterozygous for BRAF^V600E allele, was infected with HIV shRNA vectors (GL3B, mtBRAF^V600E, and wtBRAF^V600V) on day 0. The cell number was determined on day 3, 6 and 9. The in vitro cell growth was significantly suppressed with HIV-mtBRAF^V600E, but not with HIV-wtBRAF^V600V. (**b**) Western blot of the proteins harvested on day 9 in (**a**). Total BRAF proteins were decreased moderately with either HIV-mtBRAF^V600E or HIV-wtBRAF^V600V, however, the phosphorylated ERK was suppressed to a greater degree with the HIV-mtBRAF^V600E than with the HIV-wtBRAF^V600V. (**c**) HIV-mtBRAF^V600E inhibits BRAF protein almost completely in A375 melanoma cells, which is homozygous for mtBRAF^V600E, but not in 1362mel cells, which lack mtBRAF^V600E, indicating the HIV-mtBRAF^V600E acts on the mtBRAF^V600E specifically. (A part of this figure is reproduced from ref. 7 with permission from Nature Publishing Group)

protein due to gene amplification (8). Lentiviral transduction of shRNA for SKP-2 reduced in vitro cell proliferation, which was associated with the significant increase of p27^Kip1 and p21^Cip1, both of which are the substrate of SKP-2-mediated proteasomal degradation. The degree of the suppression of cell growth was correlated with the degree of SKP-2 depletion, increase of p27^Kip1 and p21^Cip1, and the increased G1 cell population (Fig. 2).

Here we will describe the method of shRNA-expressing HIV lentiviral vector production, including design of shRNA oligonucleotide templates, subcloning of the shRNA templates into the vector, HIV vector production, and vector titration.

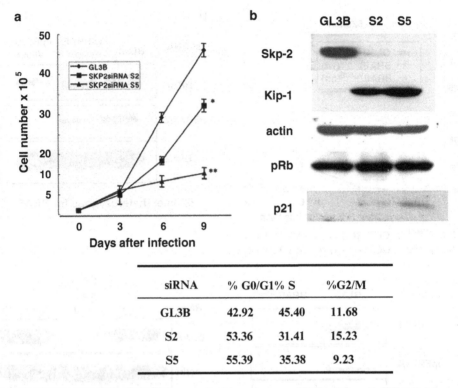

siRNA	% G0/G1	% S	%G2/M
GL3B	42.92	45.40	11.68
S2	53.36	31.41	15.23
S5	55.39	35.38	9.23

Fig. 2. Inhibition of in vitro cell growth of a human small cell lung carcinoma (SCLC) cell line with SKP-2 depletion associated with the increase of CDK inhibitors, p27^{Kip1} and p21^{Cip1}. (**a**) ACC-LC-172, a human SCLC cell line with overexpression of SKP-2 protein was infected with HIV-shRNA vectors (control GL3B, anti-SKP2 S2 and S5) at 100 moi on day 0. The cell numbers were determined on day 3, 6, and 9. The vertical bars indicate the S.D. of the triplicate assays. (* $p=0.0005$, ** $p<0.0001$). (**b**) Western blot analysis on the protein harvested on day 9. The SKP-2 protein was significantly decreased in cells infected with S2 and S5 compared to GL3B, and p27^{Kip1} and p21^{Cip1} proteins were reciprocally increased. The degrees of the increase of p27^{Kip1} and p21^{Cip1} proteins were correlated with the degrees of the decrease of the SKP-2 protein levels, as well as the increase of G0/G1 populations (*table*). The transduction efficiency (% of GFP positive cells) was equivalent among the three groups (98.7–99.9%) at the harvest. (Reproduced from ref. 8 with permission from Nature Publishing Group.)

2. Materials

2.1. HIV Vector Production and Titration

1. 293 T cells cultured in Dulbecco's Modified Eagles's Medium (DMEM) (Invitrogen, Carlsbad, CA) supplemented with 10% fetal bovine serum (FBS), penicillin and streptomycin.

2. Phosphate-buffered saline (PBS) supplemented with 0.02% (w/v) EDTA-4Na.

3. Transfer vector plasmid: HIV-U6i-GFP with an shRNA expression cassette (Fig. 3).

4. Packaging plasmid: pCMV-VSV-G-RSV-Rev and pCAG-HIVgp (both kindly provided by Dr. H. Miyoshi, BioResource Center, RIKEN Tsukuba Institute, Japan).

5. Sense and antisense oligonucleotide templates for short hairpin RNA (Invitrogen).

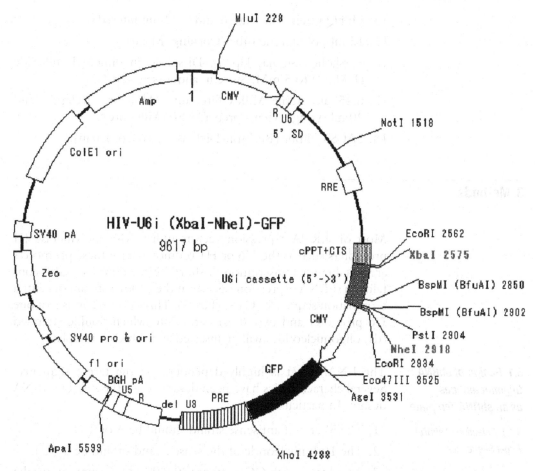

Fig. 3. The structure of HIV-U6i (*Xbal-Nhel*)-GFP vector plasmid. The HIV-U6i-GFP contains shRNA expression cassette (U6i cassette) consisting of human U6 promoter and two *BspM* (*BfuA*I) cloning sites for an shRNA template oligonucleotide, between the cPPT/CTS and the internal CMV promoter. The shRNA cassette can be excised with either *EcoR*I or *Xbal/Nhel* restriction enzymes. Multiple shRNA cassettes can be created by using *Xbal* and *Nhel* sites (*see* the text in detail). GFP gene can be expressed by the internal CMV promoter, which allows monitoring of the transduction efficiency of the vector. The viral promoter sequence in the 3′-LTR was deleted (self-inactivating vector). *RRE* Rev-responsive element, *cPPT* central polypurine tract, *CTS* central termination sequence, *CMV* cytomegalovirus promoter, *GFP* green fluorescent protein, *PRE* Woodchuck hepatitis virus posttranscriptional regulatory element

6. 1 M NaCl.

7. Thermal cycler (iCycler, Bio-Rad, Tokyo, Japan).

8. *BfuA*I restriction enzyme.

9. Ligation kit Version 2 (Takara, Kyoto, Japan).

10. DH5α competent cells (Toyobo, Osaka, Japan).

11. 14 cm culture dish (Nalgen Nunc, Rosklide, Denmark), 9 cm collagen-coated dish (Sumillon CellTight, Sumitomo Bakelite, Tokyo, Japan).

12. 2.5 M CaCl$_2$.

13. 2× HBSP: 280 mM NaCl/10 mM KCl/1.5 mM Na$_2$HPO$_4$/ 12 mM Glucose/50 mM HEPES, pH 7.12. (see Note 2).

14. MilliQ water (autoclaved and 0.22 μm filtered).

15. 15 ml-polysterene tube (Corning, Mexico).

16. Forskolin (Sigma, Japan). Dissolved in dimethyl sulfoxide (DMSO) to 5 mM. Store in aliquots at –20°C.

17. 0.45 μm sterile Millex filter unit (Millipore, Ireland), and 20 ml syringe. Centriprep YM-50 (Millipore).

18. FACS Calibur (Beckton-Dickinson, Tokyo, Japan).

3. Methods

Most of shRNA-expression vectors utilize RNA polymerase III promoters such as the U6 or H1 promoter since these promoters are suitable for transcription of short RNAs. HIV-U6i-GFP contains an shRNA expression cassette at the 3′ downstream of central termination sequence (CTS) (Fig. 3). This cassette consists of one U6 promoter and two *BfuA*I sites, into which double stranded (ds) oligonucleotides will be inserted as an shRNA template.

3.1. Design of shRNA Oligonucleotides as an shRNA Template

The RNAi effect is highly dependent on the target sequence. Several characteristics have been described for the suitable siRNA design. In particular:

3.1.1. Selection of shRNA Target Sequences

1. The 5′-end of antisense strand should be A or T (U).

2. The 10th nucleotide of the sense strand should be T (U).

3. Avoid too much GC content (30–60% of GC content may be the best).

4. The 5′-end of the sense strand must be G or A, which is a transcription start site of U6 promoter.

5. Avoid TTTT or TTATT sequence within the sense oligonucleotide, since these are the stop sequences of U6 promoter.

6. By applying to BLAST search, avoid sequences having homology to different target RNAs, which may result in the off-target effect (see Note 3). It is important especially for family of related proteins. Several algorithms are available for the siRNA target sequences. Choose four or more shRNA target sequences with high scores according to the algorithm you use.

3.1.2. Designing of shRNA Oligonucleotide Templates (Fig. 4)

1. Select appropriate linker sequence (e.g. TTCAAGAGA, or ACGTGTGCTGTCCGT); linker sequence is important for the export of shRNA from nucleus to cytoplasm, which is mediated by the binding of exportin-5 to the linker sequence.

2. The length of siRNA should be 19 nt (with 2 nt 3′-overhang). However, longer siRNAs, at most 30 nt, will also work.

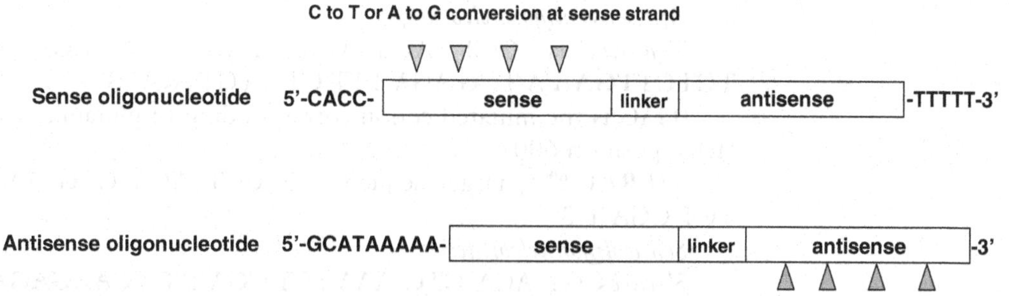

C to T or A to G conversion at sense strand

Sense oligonucleotide   5'-CACC- | sense | linker | antisense | -TTTTT-3'

Antisense oligonucleotide   5'-GCATAAAAA- | sense | linker | antisense | -3'

Fig. 4. Structure of sense and anti-sense oligonucleotide for shRNA template. Sense and anti-sense sequence of the target sequence is linked via a small linker sequence. TTTTT, a stop sequence of U6 promoter is added at the 3'-end of the sense oligonucleotide. Two to four mutations from C to T or A to G could be created on the sense strand of the sense oligonucleotide, which is useful for the stabilization of shRNA (*see* the text for details). The anti-sense oligonucleotide should be complementary for the sense oligonucleotide except for the 5'-protruding end. The 5'-protruding ends of both sense (CACC) and anti-sense oligonucleotide (GCAT) is complementary for the cloning sites of two *BfuAl* sites (*see* Fig. 3)

3. (Optional) Conversion of two to four nucleotides (from C to T or from A to G) within the sense strand of shRNA template will result in decrease of the frequency of shRNA mutation during the subcloning, and will make it easy to sequence the shRNA template (6) (see Note 4). It is important to not introduce the conversion within the antisense strand of shRNA which will affect RNAi effects significantly. Be careful to avoid creating the stop sequence (TTTT or TTATT) in the sense oligonucleotide following the nucleotide conversions.

4. Add four nucleotides 5'-overhang in both the sense and antisense oligonucleotides compatible for the cloning sites of HIV-U6i-GFP plasmid: CACC at the 5'-end of sense oligonucleotide, and GCAT at the 5'-end of antisense oligonucleotide, respectively (see Note 5).

5. Add five T (TTTTT) at the 3'-end of the sense oligonucleotide as a stop sequence (see Note 6).

6. Design the antisense oligonucleotide complementary to the sense oligonucleotide except the 5'-overhangs.

3.1.3. Examples for shRNA Oligonucleotides

These are the examples of shRNA oligonucleotides for BRAF (7) and SKP-2 (8). In these examples, no mutation is introduced in the sense strand. Bold characters indicate the linker sequence. Italic characters indicate the 5'-ends compatible for the subcloning sites in the vector.

mtBRAF[V600E]; (see Note 7) target sequence: 5'-GCT ACA GAG AAA TCT CGA T-3'

Sense oligonucleotide;

5'-*cacc* GCT ACA GAG AAA TCT CGA T **TTCAAGAGA** A TCG AGA TTT CTC TGT AGC ttttt-3'

Antisense oligonucleotide;

5'-*gcata*aaaa GCT ACA G<u>A</u>G AAA TCT CGA T **TCTCTTGAA** A TCG AGA TTT CTC TGT AGC-3'

(G<u>A</u>G is the mutated codon corresponding to glutamic acid (E) at position 600)

wtBRAF^{V600V}; target sequence: 5'-GCT ACA G<u>T</u>G AAA TCT CGA T-3'

Sense oligonucleotide;

5'-*cacc* GCT ACA G<u>T</u>G AAA TCT CGA T **T TCAAGAGA** A TCG AGA TTT C<u>A</u>C TGT AGC ttttt-3'

Antisense oligonucleotide;

5'-*gcata*aaaa GCT ACA G<u>T</u>G AAA TCT CGA T **TCTCTTGAA** A TCG AGA TTT C<u>A</u>C TGT AGC-3'

(G<u>T</u>G is the wild type codon corresponding to valine (V) at position 600)

SKP-2S2; target sequence: 5'-ATC AGA TCT CTC TAC TTT A-3'

Sense oligonucleotide;

5'-*cacc* ATC AGA TCT CTC TAC TTT A **TTCAAGAGA** T AAA GTA GAG AGA TCT GAT ttttt-3'

Antisense oligonucleotide;

5'-*gcata*aaaa ATC AGA TCT CTC TAC TTT A **TCTCTTGAA** T AAA GTA GAG AGA TCT GAT-3'

SKP-2S5; target sequence: 5'-AGG TCT CTG GTG TTT GTA A-3'

Sense oligonucleotide;

5'-*cacc* AGG TCT CTG GTG TTT GTA A **TTCAAGAGA** T TAC AAA CAC CAG AGA CCT ttttt-3'

Antisense oligonucleotide;

5'-*gcata*aaaa AGG TCT CTG GTG TTT GTA A **TCTCTTGAA** T TAC AAA CAC CAG AGA CCT-3'

GL3B (anti-firefly luciferase; as a control shRNA); target sequence: 5'-GTG CGC TGC TGG TGC CAA C-3'

Sense oligonucleotide;

5'-*cacc* GTG CGC TGC TGG TGC CAA C **TTCAAGAGA** G TTG GCA CCA GCA GCG CAC ttttt-3'

Antisense oligonucleotide;

5'-*gcata*aaaa GTG CGC TGC TGG TGC CAA C **TCTCTTGAA** G TTG GCA CCA GCA GCG CAC-3'

3.2. In vitro Annealing of Sense and Antisense Oligonucleotides

1. Prepare the following mixture:

Sense oligonucleotide	(100 µM)	5 µl
Antisense oligonucleotide	(100 µM)	5 µl
1 M NaCl		2 µl
H₂O		8 µl (total 20 µl)

2. By using thermal cycler, make double-stranded (ds) oligonucleotide by in vitro annealing reaction as follows: 99°C for 2 min, cool down to 72°C, and cooling down slowly to 4°C for 2 h.

3.3. Subcloning ds Oligonucleotides into the HIV-U6i-GFP

1. To prepare BfuAI-digested HIV-U6i-GFP fragment for the ligation to the ds oligonucleotide, digest HIV-U6i-GFP plasmid with BfuAI restriction enzyme.

2. Apply to agarose gel electrophoresis.

3. Recover the digested plasmid from agarose gel by using a commercial kit.

4. For the ligation reaction, prepare:

*Bfu*AI-digested HIV-U6i-GFP	0.03 pmol (176 ng)
1:100 diluted ds oligonucleotide	1 µl (0.25 pmol)
Solution I of the TAKARA Ligation Kit Ver.2	One volume of the above mix (see Note 8).

5. Incubate at 16°C for more than 2 h or overnight.

6. Transform 10 µl of DH5α or other appropriate competent cells with 1–2 µl of the ligation mix.

7. Pick up 6 or more colonies of the transformed DH5α, amplify them in a small scale (~2 ml) and purify the plasmids.

8. By using the vector sequence 5′-upstream of the shRNA cassette as a sequence primer (i.e., AGCAACAGACATA-CAAACTAAAGA), sequence the plasmids to confirm the correct subcloning of the ds oligonucleotide (see Note 9).

3.4. Multiple shRNA Cassette

In HIV-U6i-GFP, the shRNA cassette is flanked by 5′-*Xba*I and 3′-*Nhe*I restriction sites (Fig. 3), which recognize T CTAGA, and G CTAGC, respectively, and the cohesive ends digested with these enzymes are complementary to each other. Therefore, an *Xba*I-*Nhe*I fragment containing one shRNA cassette in one HIV shRNA vector can be subcloned into the *Xba*I or *Nhe*I site of another HIV shRNA vector (9–11). In the middle point of the tandemly aligned shRNA cassettes, ligation between XbaI and NheI sites will diminish both sites, leaving one 5′-*Xba*I and one 3′-*Nhe*I site, respectively. You can arrange multiple shRNA cassettes by repeating this step within the capacity of recombinant viral genome (up to 10 or more shRNA cassettes). With this method, multiple RNAi can be attained with one HIV shRNA vector (9–11).

3.5. Production of-shRNA Expressing Lentiviral Vector

The method of HIV-shRNA vector production is the same as the usual HIV lentiviral vector production protocol. Here, we describe our protocol shortly.

1. Prepare 293 T cells in an exponential growth phase, and confluent in 14 cm-dish at the harvest.

2. 293 T cell harvesting: Change the medium to PBS/0.02% EDTA, and incubate at 37°C for 5 min.

3. Detach the cells from the plate by vigorous pipetting, collect them by centrifugation.

4. Dispense the single cells at 1:5 into 9 cm-collagen-coated dish (*see* Note 10).

5. Incubate at 37°C, 5% CO_2 more than 30 min.

6. Prepare the DNA-Ca^{2+} mix as follows;

Transfer vector (HIV-shRNA plasmid)		17 µg
pCMV-VSV-G-RSV-Rev (VSV-G and Rev)		10 µg
pCAG-HIVgp (gag-pol)		10 µg
H_2O	Total	450 µl (per one dish)

7. Add 50 µl of 2.5 M $CaCl_2$ (total 500 µl), mix well by vortexing.

8. Add the above DNA-Ca^{2+} mix slowly into one volume of 2× HBSP (pH 7.12) in a polysterene tube, mix well. Incubate at room temperature for 30 min.

9. Pour the DNA-Ca2+mix precipitates onto the 293 T cell as described in step 3.

10. Incubate at 37°C, 5%CO_2 12–16 h.

11. Replace the medium with fresh complete medium (DMEM/10% FBS) (7–10 ml per dish), then add forskolin (see Note 11) at a final concentration of 10 µM. Incubate for further 48 h.

12. Collect the culture supernatant, filtrate through 0.45 µm sterile Millex filter unit to remove the floating cells.

13. Concentrate the lentiviral vector preparation by ultrafiltration using Centriprep YM-50 (Centrifugal filter device) (see Note 12). The final virus volume will be 500–800 µl. Snap frozen the aliquots and keep at –85°C until use.

3.6. Titration of-shRNA Expressing Lentiviral Vector

Since HIV-U6i-GFP encodes green fluorescent protein (GFP) driven from internal CMV promoter (Fig. 3), you can titrate the virus by evaluating the GFP expression. Here, we describe our method.

1. Dispense 293 T cells at 10^5 cells/well in a 6-well plate the day before HIV-1 vector infection (day - 1).

2. Challenge the 293 T cells with the concentrated HIV-shRNA lentiviral vector at 0, 0.5, 1, 2 and 5 µl/well.

3. Incubate at 37°C, 5% CO_2 for 3 days (day 0).

4. Harvest the 293 T cells using PBS(-)/0.02%EDTA. Collect the cells and fix them in 2% paraformaldehyde.

5. Evaluate the percentage of GFP-positive cells by flow cytometry (day 3).

6. Calculate the lentiviral vector titer assuming that 4×10^5 cells were infected by the lentiviral vector preparation.

   e.g. If 30% of 293 T cells were GFP-positive with 1 μl of virus infection, $(4 \times 10^5 \text{ cells} \times 0.30) \div (1 \times 10^{-3} \text{ml}) = 1.2 \times 10^8$ TU (transducing unit)/ml.

4. Notes

1. In general, the use of siRNAs less than 30 base pairs could circumvent strong interferon (IFN) responses. However, it has been revealed that IFN responses cannot be bypassed even with siRNAs completely (12). Strong, sequence-dependent IFN responses may be observed via TLR7 stimulation in plasmacytoid dendritic cells (PDCs) (13).

2. The pH of 2× HBSP is crucial for the production of fine precipitates.

3. Off-target effect is nonspecific silencing of irrelevant genes with partial homology to the target sequence of siRNA (shRNA). Avoid target sequences with at least 3 mismatches with any irrelevant genes. It is important to confirm phenotypic changes after RNAi with more than two different target sequences to the same gene to exclude the off-target effect.

4. The change from C to T or from A to G in the sense strand of shRNA template without change in the complementary antisense strand will result in the change from C:G to U:G paring, or from A:U to G:U paring within the short hairpin RNA, which can maintain the duplex structure. These nucleotide conversions will make the DNA template structure more stable for subcloning or sequencing (5).

5. BfuAI (BspMI) is a type II endonuclease. BfuAI recognizes ACCTGC sequence, and digest the 4th nucleotide 3'-downstream of this sequence, creating a 5'-protruding end. Therefore, these four nucleotides are compatible for subcloning in this particular plasmid vector.

6. TTTT is a usual stop sequence of the U6 promoter. Here, we introduce five Ts to avoid any potential failure of the transcription stop.

7. The shRNA targets for both wild-type (wt) and mutated (mt) BRAF are shown. mtBRAF^{V600E} is specific for glutamic

acid codon (G<u>A</u>G), while wt BRAF^{V600V} is specific for valine codon (G<u>T</u>G).

8. It is important to keep the total volume less than 10 µl in this recipe for efficient ligation reaction.

9. Introduction of two to four mutations in the sense strand (see Note 4) could make the sequencing of the shRNA template easy. Without the sense strand mutation, G:C pairings between the sense and antisense strands make the DNA template a firm, stem-like structure, and the sequence reaction will often stop in the hairpin structure.

10. The area of 14 cm dish is about 2.4 times of 9 cm dish. When you passage the confluent cells on 14 cm dish at 1:5 to 9 cm dish, the cells will occupy more than half area of the dish.

11. Forskolin activates adenylate cyclase, leading activation of protein kinase A via cAMP, which results in the increase of CMV promoter activity, hence increasing the viral titer.

12. Ultracentrifugation method may be usually selected for the concentration of HIV vectors. The method you choose depends on the scale of the experiment. Ultracentrifugation method is suitable for the larger scale preparation.

Acknowledgments

The authors thank Dr. Miyagishi, M. for providing the shRNA cassettecontaining plasmid (pU6i) and for his helpful advice, and also Dr. Miyoshi, H. for providing us the HIV-1 vector system.

References

1. Fire, A., Xu, S., Montogeomery, M.K., Kostas, S.A., Driver, S.E., and Mello, C.C. (1998) Potent and specific genetic interference by double-stranded RNA in caenorhabditis elegans. *Nature*. **391**, 806–11.

2. Hammond, S.M., Bernstein, E., Beach, D., and Hannon, G. (2000) An RNA-directed nuclease mediates post-transcriptional gene silencing in Drosophila cells. *Nature*. **404**, 293–6.

3. Elbashir, S. M., Harborth, J., Lendeckel, W., Yalcin, A., Weber, K., and Tuschl, T. (2001) Duplexes of 21-nucleotide RNAs mediate RNA interference in cultured mammalian cells. *Nature*. **411**, 494–8.

4. Brummelkamp, T. R., Bernards, R., and Agami, R. (2002) A system for stable expression of short interfering RNAs in mammalian cells. *Science*. **296**, 550–3.

5. Yi, R., Qin, Y., Macara, I. G., and Cullen, B. R. (2003) Exportin-5 mediates the nuclear export of pre-microRNAs and short hairpin RNAs. *Genes Dev*. **17**, 3011–6.

6. Miyagishi, M., Sumimoto, H., Miyoshi, H., Kawakami, Y., and Taira, K. (2004) Optimization of an siRNA-expression system with an improved hairpin and its significant suppressive effects in mammalian cells. *J. Gene Med*. **6**, 715–23.

7. Sumimoto, H., Miyagishi, M., Miyoshi, H., Yamagata, S., Shimizu, A., Taira, K., and Kawakami, Y. (2004) Inhibition of growth and invasive ability of melanoma by inactivation of mutated BRAF with lentivirus-mediated RNA interference. *Oncogene*. **23**, 6031–9.

8. Sumimoto, H., Yamagata, S., Shimizu, A., Miyoshi, H., Mizuguchi, H., Hayakawa, T.,

Miyagishi, M., Taira, K., and Kawakami, Y. (2005) Gene therapy for human small-cell lung carcinoma by inactivation of Skp-2 with virally mediated RNA interference. *Gene Ther.* **12**, 95–100.

9. Sumimoto, H., Hirata, K., Yamagata, S., Miyoshi, H., Miyagishi, M., Taira, K., and Kawakami, Y. (2006) Effective inhibition of cell growth and invasion of melanoma by combined suppression of BRAF (V599E) and Skp2 with lentiviral RNAi. *Int. J. Cancer.* **118**, 472–6.

10. Sumimoto, H., Imabayashi, F., Iwata, T., and Kawakami, Y. (2006) The BRAF-MAPK signaling pathway is essential for cancer immune evasion in human melanoma cells. *J Exp. Med.* **203**, 1651–56.

11. Sumimoto, H. and Kawakami, Y. (2007) Lentiviral vector-mediated RNAi and its use for cancer research. *Future Oncol.* **3**, 655–64.

12. Sledz, C.A., Holko, M., de Veer, M.J., Silverman, R.H., and Williams, B.R.G. (2003) Activation of the interferon system by short-interfering RNAs. *Nat. Cell Biol.* **5**, 834–9.

13. Hornung, V., Guenthner-Biller, M., Bourquin, C., Ablasser, A., Schlee, M., Uematsu, S., Noronha, A., Manoharan, M., Akira, S., de Fougerolles, A., Endres, S. and Hartmann G. (2005) Sequence-specific potent induction of INF-α by short interfering RNA in plasmacytoid dendritic cells through TLR7. *Nat. Med.* **11**, 263–70.

Chapter 14

Lentiviral Vector Engineering for Anti-HIV RNAi Gene Therapy

Olivier ter Brake, Jan-Tinus Westerink, and Ben Berkhout

Abstract

RNA interference or RNAi-based gene therapy for the treatment of HIV-1 infection has recently emerged as a highly effective antiviral approach. The lentiviral vector system is a good candidate for the expression of antiviral short hairpin RNAs (shRNA) in HIV-susceptible cells. However, this strategy can give rise to vector problems because the anti-HIV shRNAs can also target the HIV-based lentiviral vector system. In addition, there may be self-targeting of the shRNA-encoding sequences within the vector RNA genome in the producer cell. The insertion of microRNA (miRNA) cassettes in the vector may introduce Drosha cleavage sites that will also result in the destruction of the vector genome during the production and/or the transduction process. Here, we describe possible solutions to these lentiviral-RNAi problems. We also describe a strategy for multiple shRNA expression to establish a combinatorial RNAi therapy.

Key words: RNAi, siRNA, shRNA, miRNA, Lentiviral titer

1. Introduction

Double-stranded RNA can induce sequence-specific inhibition of a target mRNA via the process known as RNA interference (RNAi). We and others have shown that this technology can be employed for efficient suppression of HIV-1 replication, either through targeting of the HIV-1 RNA genome or the mRNAs for cellular cofactors that are essential for HIV-1 replication. As with conventional drugs, it has been demonstrated that a single RNAi inhibitor is not sufficient to durably inhibit HIV-1 replication, and virus escape variants were selected with a mutation in the target sequence. Thus, multiple RNAi inhibitors have to be used simultaneously to prevent viral escape (1, 2). The therapy of a chronic virus infection will require a gene therapy approach with

Maurizio Federico (ed.), *Lentivirus Gene Engineering Protocols*, Methods in Molecular Biology, vol. 614,
DOI 10.1007/978-1-60761-533-0_14, © Humana Press, a part of Springer Science+Business Media, LLC 2010

a vector system that provides durable inhibition. For this, the lentiviral vector is ideally suited because it stably integrates into the host cell genome. In a straightforward approach, several labs have inserted multiple repeats of the same shRNA expression cassette within the lentivector genome, which generates repeats of the promoter that controls the expression of the different shRNAs (3, 4). However, the presence of sequence repeats within the lentiviral vector genome gives rise to vector instability due to recombination and deletion (5). This problem can be solved by using different promoters for shRNA expression, thus avoiding repeat sequences. We used three RNA polymerase III cassettes and one polymerase II cassette. Here we describe the cloning strategy for such a multiple shRNA expressing lentiviral vector.

We incorporated a quadruple-shRNA cassette within the lentiviral vector genome in the antisense orientation (Fig. 1). This design yielded a greatly reduced transduction titer (approximately 18-fold). Insertion of the shRNA cassettes in the sense orientation was not chosen because putative transcription termination sites will be introduced in the vector genome. In fact, polymerase III termination sites have been shown to be recognized by polymerase II (6). We have tested viral vectors that contain each of the cassettes in a singular fashion, and found that vector production is only marginally reduced (approximately twofold). This indicates that the introduced sequences per se do not cause the reduction in vector titer (7). It is unlikely that the antisense oriented polymerase III units will interfere with sense transcription of the lentiviral genome, and the results with the singular cassettes indicate that even the U1 polymerase II unit does not cause polymerase collision (8). Rather, the titer decrease can be the consequence of the increase in vector genome size (9). Alternatively, the introduction of multiple relatively stable hairpin structures may negatively affect the lentiviral transduction titer.

We have analyzed five possible routes through which shRNAs against HIV-1 may affect lentiviral vector production (Fig. 2) (10). All shRNAs used in this analysis were demonstrated to be potent inhibitors of HIV-1 replication and reporter gene expression (Fig. 3) (11). The lentiviral vector RNA genome is potentially an efficient target for RNAi-mediated degradation during vector production. We tested several approaches to overcome these problems. The most general solution to restore vector titers is by saturation of the RNAi pathway in the producer cell with siRNAs or shRNAs (Fig. 4). This saturation strategy can be of general use in solving lentiviral titer problems, for instance with miRNA expression constructs. Interestingly, in contrast to the native HIV-1 sequences within the lentiviral vector genome, the shRNA-encoding sequence itself is not a target for RNAi-mediated attack. The absence of such self-targeting is likely caused by the fact that the target sequence is embedded within a stable

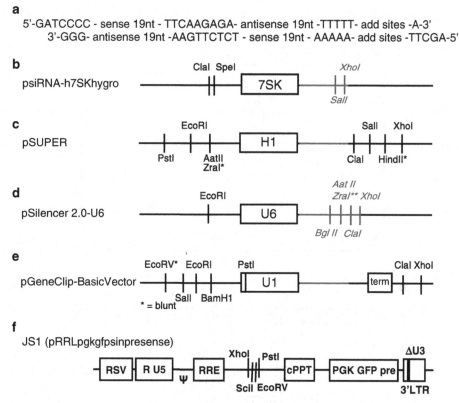

a

5'-GATCCCC - sense 19nt - TTCAAGAGA- antisense 19nt -TTTTT- add sites -A-3'
3'-GGG- antisense 19nt -AAGTTCTCT - sense 19nt - AAAAA- add sites -TTCGA-5'

Fig. 1. Cloning strategy for a lentiviral vector with multiple shRNA cassettes. (**a**) Schematic of the shRNA-encoding unit made from annealed oligonucleotides and designed for insertion into the H1 polymerase III expression cassette of pSUPER. Appropriate overhangs are created to enable the orientation-specific ligation behind the H1 promoter. The 19-nucleotide sense strand is linked via a 9-nucleotide loop to the antisense strand of the hairpin, which is terminated by the T5 termination signal. Additional restriction sites can be incorporated downstream of the T5 signal to facilitate the cloning strategy. (**b–f**) Cloning strategy for the 4-shRNA lentiviral vector. Indicated are restriction sites of the original constructs, while indicated in grey/italics are the sites that were added to facilitate cloning. First, the 7SK shRNA cassette (**b**) is inserted behind the H1 cassette (**c**) using the ClaI and SalI restriction sites. The U6 shRNA cassette (**d**) is inserted in front of the U1 cassette (**e**) using the EcoRI/BglII and EcoRI/BamHI sites, respectively. Subsequently, the U6/U1 double cassette is excised using EcoRV/XhoI and inserted behind the H1/7SK cassettes with HindII/XhoI, resulting in the 4-shRNA expression construct. Finally, the 4-shRNA cassettes were inserted in the lentiviral vector backbone (**f**). We used the third-generation self-inactivating lentiviral vector JS1. The vector genome is expressed from the Rous Sarcoma Virus promoter (RSV) and contains the Rev-responsive element (RRE), the central polypurine tract (cPPT), and a "self-inactivating" deletion in the U3 region (ΔU3) of the 3' LTR. The enhanced green fluorescent protein (GFP) reporter gene is expressed from the phosphoglycerate kinase promoter (PGK). This transcript includes the posttranscriptional regulatory element (pre) from hepatitis B virus. Note that U1-mediated shRNA expression uses polymerase II that terminates 10-nucleotides upstream of the specific 14-nucleotide termination signal. In case one needs to incorporate additional restriction sites, these $10+14=24$ nucleotides need to be part of the insert design (e.g., exchange the T5 sequence of 1A with these 24 nucleotides)

RNA structure that renders the sequence inaccessible to the RNA-induced silencing complex (RISC) (12).

Targeting of the Gag-Pol construct reduces capsid production and the transduction titer, but this problem can simply be overcome by switching to a human codon-optimized Gag-Pol

204

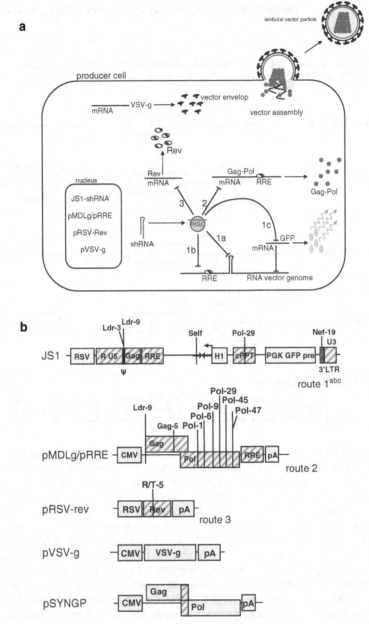

Fig. 2. Lentiviral vectors carrying anti-HIV shRNAs: RNAi against vector components that may reduce vector titer during production. (**a**) Schematic of potential shRNA targets during lentiviral vector production. The expressed shRNA is processed and incorporated into RISC. This activated RISC can target the vector genome via three routes. Route 1a targets the shRNA sequence as part of the lentiviral vector genome (self-targeting). Route 1b targets the vector genome. Route 1c: when the 3'LTR of the vector genome is targeted, the reporter gene is also targeted. The Gag-Pol, route 2, and the Rev transcript, route 3, can also be targeted. (**b**) Plasmids used for lentiviral vector production. Target sites for effective shRNAs are indicated and HIV-1 sequences are striped. Expression cassettes for shRNAs under control of the H1 promoter were inserted in the lentiviral vector genome. Any shRNA has the potential to target its own sequence as part of the lentiviral vector genome (self-targeting). The vector transcript is expressed from the RSV promoter and starts with the viral R and U5 regions, the packaging signal (Ψ) and part of the Gag open reading frame (Gag). It also contains the Rev-responsive element (RRE), central polypurine tract (cPPT), and self-inactivating 3' LTR (see legend to Fig. 1 for further details). Transcription of both the vector genome and the GFP reporter terminates at the HIV-1 polyA signal within the 3' LTR. The HIV-1 Gag-Pol gene was expressed from the pMDLg/pRev plasmid with the RRE. Alternatively, we used pSYNGP with a human-codon optimized sequence without RRE. The Rev protein was expressed from pRSV-rev [5] and was included in all vector production schemes. The viral vector was pseudo-typed with the VSV-g protein. Figure reprinted from ter Brake et al. 2007 with permission (10)

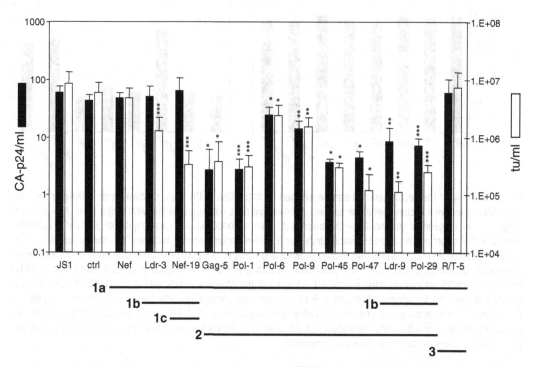

Fig. 3. HIV-1 sequences within the vector genome and the HIV-1 Gag-Pol transcript are efficient RNAi targets during vector production. CA-p24 production (*black bars*, left *y*-axis) and transduction titer (*white bars*, right *y*-axis) of the different lentiviral vector stocks (*x*-axis) were determined. Indicated below the x-axis are the routes through which lentiviral vector production can be targeted. RNAi against HIV-1 sequences within the vector backbone (1b and 1c) results in a reduced transduction titer while capsid production was unaffected. The notable exception is self-targeting of the shRNA encoding sequence (route 1a), which does not affect the transduction titer as the vector titers are the same as the inactive control shRNA (ctrl). Targeting of Gag-Pol mRNA results both in reduced capsid and transduction titers (route 2). When Rev mRNA is targeted (route 3), no significant reduction in capsid production and transduction titer was observed. Average values are shown with the error bars indicating the standard deviations, and p-values relative to JS1 are indicated (*$p = 0.05$; **$p = 0.03$; ***$p = 0.01$; $N = 2-5$). Figure reprinted from ter Brake et al. 2007 with permission (10)

construct (Fig. 5). Finally, we measured no effect with a shRNA that targets the Rev mRNA, which likely indicates that sufficient Rev is expressed - despite the RNAi-inhibition – to execute its functions during lentivector production.

2. Materials

2.1. shRNA Vector Construction

1. Annealing buffer (100 mM NaCl and 50 mM HEPES pH 7.4).

2. Bacterial strain GT116 (Invivogen, San Diego, CA).

3. BigDye Terminator v3.1 Cycle Sequencing Kit, containing 1 M Betaine (Sigma, St Louis, MO).

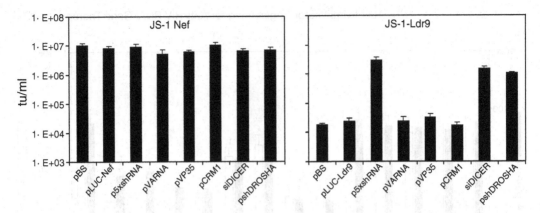

Fig. 4. Targeting of the vector genome can be avoided by saturation of the RNAi pathway with siRNAs or shRNAs. We produced lentiviral vectors that either express JS1-Nef that does not target the lentiviral vector (**a**) or JS1-Ldr9 that targets the vector genome and the Gag-Pol transcript (**b**). Vector production was carried out with pSYNGP to avoid the latter Gag-Pol targeting. We tested several strategies to solve vector genome targeting as marked on the x-axis. Transduction titer of the control pBS transfection shows a 3-log drop in JS1-Ldr9 production compared to JS1-Nef. Using decoy RNAi targets (pLUC-Ldr9), an RNAi suppressor VA RNA (pVARNA) or VP35 protein (pVP35), and CRM1 (pCRM1) overexpression did not improve titer. In contrast, silencing of the RNAi machinery with siDicer (100 nM) or shDrosha greatly improved the transduction titer, and shRNA overexpression from p5xshRNA boosted the titer even further to normal levels. Although the siDicer and shDrosha effect can possibly be attributed to knockdown of Dicer and Drosha, we realize that it could also reflect RISC saturation. Consistent with this idea is the competing effect of any siRNA/shRNA that prevents Ldr9-mediated targeting of the vector genome

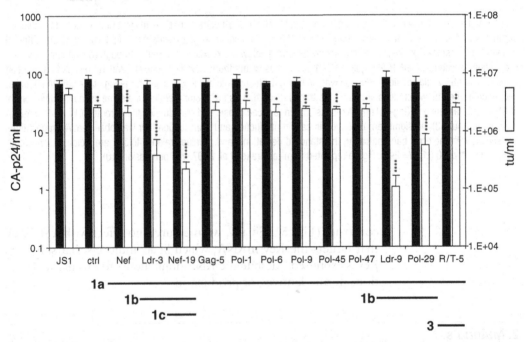

Fig. 5. Targeting of the HIV-1 Gag-Pol transcript is avoided using a synthetic human codon optimized transcript. Lentiviral vector production with a human codon-optimized Gag-Pol transcript. CA-p24 production (black *bars*, left y-axis) and transduction titer (*white bars*, right y-axis) of the different lentiviral vector stocks (x-axis) were determined. CA-p24 production was restored in all vector productions since RNAi route 2 was abrogated. Targeting of HIV-1 sequences in the vector backbone remains possible in this context. Indicated below the x-axis are the routes through which lentiviral vector production can be targeted. Averages values are shown with the error bars indicating the standard deviations, and p-values relative to JS1 are indicated (*$p = 0.12$; **$p = 0.10$; ***$p = 0.05$;****$p = 0.03$; *****$p = 0.01$; $N = 2$–5). Figure reprinted from ter Brake et al. 2007 with permission (10)

2.2. Cell Cultures

1. Dulbecco's Modified Eagle's Medium (DMEM) (Gibco/BRL, Bethesda, MD) supplemented with 10% fetal bovine serum (FCS, Hyclone, Ogden, UT), penicillin (100 U/ml) and streptomycin (100 µg/ml).

2. Advanced Roswell Park Memorial Institute (RPMI) medium (Gibco/BRL) supplemented with 2 mM L-glutamine, 1% FCS, 40 U/ml penicillin and 40 µg/ml streptomycin.

3. Phosphate Buffered Saline solution (PBS), pH 7.4 (Gibco/BRL).

4. 0.05% Trypsin with EDTA solution (Gibco/BRL).

5. Human Embryonic Kidney 293 T cell line (ATCC nr CRL-11268).

6. SupT1 T cell line (ATCC nr CRL-1942).

2.3. Constructs for Lentiviral Vector Production

1. Lentiviral vector pRRLcpptpgkgfppreSsin (13), which was renamed JS1.

2. shRNA expression constructs with the polymerase III promoters human H1 (14), U6 (15) and 7SK (16), and the human U1 polymerase II promoter (17) are commercially available. We used the pSUPER (OligoEngine, Seattle, WA, USA), pSilencer 2.0-U6 (Ambion Inc., Austion TX, USA), psiRNA-h7SKhygro (Invivogen), and the pGeneClip-BasicVector (Promega Corp., Madison, WI) constructs that contain the H1, U6, 7SK, and U1 promoters, respectively.

3. Lentiviral vector packaging constructs. For expression of the HIV-1 gag-pol genes, we used pMDLg/pRRE (18) or pSYNGP (19). pMDLg/pRRE contains the wild-type HIV-1 sequence, and pSYNGP expresses a human codon-optimized gag-pol sequence. Only pMDLg/pRRE encodes the RRE. pRSV-rev (18) was used for Rev expression.

4. pVSV-g was used for expression of the Vesicular Stomatitis virus (VSV)-g envelope protein (18).

5. Luciferase reporters with an RNAi target were used as decoy: pLUC-Nef and pLUC-Ldr9 (11).

6. The pSUPER construct p5xshRNAcontains five repeated H1 promoters (11) that express shRNAs against the following HIV-1 targets: Gag-5, Pol-1, Pol-6, Pol-9, and Pol-47. This plasmid was used to provide excess shRNAs for saturation experiments.

7. pVARNA is an expression construct that synthesizes Adenovirus VA RNA transcripts (20), which are potent RNAi suppressors.

8. pVA35 expresses the Ebola VP35 protein (21–23), which is an RNAi suppressor protein.

9. pCRM1 expresses the CRM1 cofactor that can increase vector genome export (24) (kindly donated by Tom Hope, Northwestern University, Chicago).

10. pshDrosha expresses an shRNA that knocks down the Drosha enzyme function (kindly donated by Bryan Cullen, Duke University, Durham).

11. siDicer is an siRNA that knocks down expression of the Dicer enzyme.

2.4. Lentiviral Vector Production

1. 70 μm Nylon cell strainers (BD Falcon, Bedford, MA).

1. Optimem medium (Gibco/BRL).

2. Lipofectamine-2000 (Invitrogen, Carlsbad, CA).

3. 0.45 μm cellulose acetate filter (Whatman, Clifton, NJ).

4. 100 kD MWCO Amicon Ultra centrifugal filter devices (Millipore, Billerica, MA).

5. Lentiviral vector construct JS-1 or its derivatives.

6. Lentiviral vector packaging constructs pMDLg/pRRE or pSYNGP, the envelope construct pVSV-g and pRSV-rev.

7. Optional: additional constructs that may specifically enhance lentiviral vector production when RNAi-related production problems are encountered.

3. Methods

3.1. shRNA Vector Construction

Figure 1 provides an outline of the multiple shRNA cloning strategy as an example.

1. The shRNA genes were made by designing complementary sense and antisense DNA oligonucleotides with appropriate overhangs and additional restriction sites to facilitate subsequent cloning. The oligonucleotides can be custom ordered. We use standard shRNA design that includes perfect complementary 19 nucleotide sense and antisense strands connected by the 9-nucleotide Brummelkamp loop design (UUCAAGAGA) of the pSuper system (14) (see Note 1).

2. Dissolve the DNA oligonucleotides in sterile H_2O to a final concentration of 3 mg/ml.

3. Dilute 1 μl of each oligonucleotide in 48 μl of annealing buffer.

4. Heat at 90°C for 4 min, cool down to 70°C for 10 min, and gradually cool further to 10°C over 30 min. The annealed oligonucleotides can either be used immediately in a ligation reaction

or they can be stored at 4°C for a short period and at −20°C for a prolonged period.

5. Ligate 1 μl of the annealed oligonucleotides with 100 ng of the appropriate vector backbone.

6. Transform the ligation mixture to a bacterial strain compatible with stable RNA hairpin structures and inverted repeat containing plasmids, for instance GT116.

7. Sequence positive colonies with a standard sequence reaction, for instance BigDye Terminator v3.1 Cycle Sequencing Kit (see Note 2). However, not all hairpin constructs can be sequenced using standard sequencing conditions. In that case, one could add 1 M Betaine to improve the quality of the sequence reaction.

8. Combine shRNA expression cassettes using standard cloning techniques and insert them in the lentiviral vector backbone.

3.2. Lentiviral Vector Production

1. In the morning of day 1, trypsinize 293 T cells and pass them through a cell strainer to remove clumps of cells. Count the cells and seed 2.5×10^6 cells for the transfection in a T25 (25 cm^2) flask in 5 ml culture medium.

2. Transfect the 293 T cells in the afternoon of day 2. Prepare the DNA–lipofectamine 2000 complexes in medium for each transfection as follows:

(a) Dilute 2.4 μg lentiviral vector plasmid, 1.5 μg pSYNGP (or pMDLg/pRRE), 0.83 μg pVSV-g, and 0.6 μg pRSV-rev in 0.75 ml Optimem. One can also include 4.7 μg of an additional plasmid (see Subheading 2.3, **items 5–10**), or 100 nM of siDICER (see Subheading 2.3, **item 11**: use 0.4 nmol in final volume of 4 ml).

(b) Mix Lipofectamine-2000 gently before use, and dilute 16 μl in 0.75 ml of Optimem, and incubate for 5 min at room temperature.

(c) Combine the DNA and Lipofectamine 2000 dilutions, and mix gently. Incubate for 20 min to allow the formation of DNA–Lipofectamine 2000 complexes.

(d) Subsequently, add 2.5 ml DMEM supplemented with 10% FCS, but without antibiotics.

3. Replace the cell medium with the prepared DNA–Lipofectamine 2000 complexes.

4. The next day, remove the transfection medium, and replace with 3 ml Optimem supplemented with penicillin (30 U/ml), streptomycin (30 μg/ml), and CaCl$_2$ (100 μg /ml) (see Notes 3 and 4).

5. Harvest the lentiviral vector on day 4. Spin down cellular debris by low speed centrifugation (238–370×g) for 10 min and subsequently filter the viral supernatant over a 0.45 μm filter. The highest titers are obtained at day 4. Optionally, additional vector can be harvested at day 5, yielding a lower titer (approximately tenfold reduction in transduction titer as compared to day 4).

6. The cleared lentiviral vector supernatants can be concentrated with a centrifugal filter (MWCO 100 kD) at 2,500×g for 30 min.

7. Aliquot the supernatants and store at –80°C.

8. Determine transduction titer and capsid titer of frozen aliquots. We routinely use the SupT1 T cell line to determine the transduction titer. The capsid CA-p24 titer can be determined with CA-p24 ELISA according to the following protocol. 96-well plates are coated with anti-HIV-1 CA-p24 antibody (D7320 Aalto Bio Reagents). Dissolve antigen in 2 ml PBS, dilute 0.5 ml in 50 ml 0.05 M $NaHCO_3$ pH 8.5 for 20 plates (25 μl per well), and incubate for 24 h at room temperature. Prepare the stock CA-p24 antigen in TBS (1% empigem, 10% sheep serum) and store at –80°C. To prepare a standard curve, the stock is diluted from 0 to 5 ng/ml in TBS + 0.05% empigem. Wash the plates 4× with TBS, apply the samples (and standard), and incubate for 2 h at room temperature, wash the plates 4× with TBS. Alkaline phosphatase conjugated antibody is diluted 40,000× in TBS with 20% sheep serum, 0.02 g/ml milk powder, and 1% Tween. Add 25 μl per well and incubate for 1 h at room temperature, wash 4× with PBS + 0.1% Tween, add 25 μl substrate, incubate for 30 min, and measure the luminescence on a luminescence reader.

3.3. Transduction Titer

1. Seed 50,000 SupT1 cells in 100 μl in the 96-well format (see Note 5). SupT1 cells are cultured in Advanced Roswell Park Memorial Institute (RPMI) medium supplemented with 2 mM L-glutamine, 1% FCS, 40 U/ml penicillin, and 40 μg/ml streptomycin.

2. Add a dilution series of lentiviral vector stock (see Note 6). We routinely use 10, 1, and 0.1 μl of the stock in an endvolume of 100 μl. Total volume in each well is now 200 μl.

3. After 6 h, carefully aspirate the medium from each well, avoiding to disturb the cells on the bottom of the flask. Round bottom wells are most handy, but flat bottom wells do also function. Replace the medium with 200 μl of SupT1 medium.

4. Carefully aspirate medium from each well at 48 h posttransduction, avoiding to disturb the cells on the bottom of the flask. Resuspend the cells in 200 μl FACS buffer (PBS with 1% FCS) with 4% paraformaldehyde and perform FACS analysis (see Notes 7 and 8).

4. Notes

1. The construction of shRNA expression cassettes with oligonucleotides allows the incorporation of flanking restriction sites, which is very handy in combinatorial RNAi strategies that use multiple shRNA cassettes.

2. Sequencing of shRNA constructs can still be problematic even when using optimized sequencing conditions. We have described that the introduction of G-U base pairs in the shRNA dramatically improves the sequencing reaction (25). Multiple G-U base pairs can in fact be introduced in the shRNA molecule without loss of RNAi-function. An alternative solution for the sequencing problem is the introduction of a restriction site within the loop, which allows run-off sequencing from both sides of the hairpin after restriction enzyme digestion of the DNA.

3. Production in Optimem without serum improves vector titer about twofold as compared to standard DMEM with 10% serum. More importantly, the absence of serum improves the filtration and concentration of supernatants. Serum increases viscosity and clogs filters, but also adds a large amount of contaminating protein.

4. $CaCl_2$ is added to Optimem during lentivirus production to improve attachment of the cells in the absence of serum.

5. For determining the transduction titer, we routinely use SupT1 cells because it enables a relative high-throughput titration of a large number of transductions. For instance, we routinely transduce SupT1 cells in the 96-well format, which avoids laborious cell trypsinization and resuspension steps. In a direct comparison using similar conditions (50,000 cells in 200 µl medium, 6 h transduction, FACS analysis 48 h posttransduction), we have found that the transduction titer obtained in SupT1 cells is about twofold higher than in 293 T cells.

6. We routinely concentrate lentiviral vector stocks 40-fold, from an initial volume of 10 ml to 250 µl. Yields vary but are usually at least 50%, with transduction titers after concentration of approximately 5×10^8 tu/ml, but occasionally as high as 1×10^9 tu/ml.

7. Importantly, we have found that the CA-p24 production is relatively constant for a wide variety of lentiviral vectors, which may differ significantly in their transduction titer.

8. Some RNAi-lentiviral vectors exhibit a more severely reduced CA-p24 production and/or transduction titer. In that case, one can use concentrated lentiviral vector stocks to reach a similar transduction titer as tested on SupT1 cells. Despite

this correction, we have observed an approximately tenfold reduction of the transduction efficiency of the 4-shRNA vector on primary CD34+ blood stem cells. There are several possible causes for this cell type specific effect. The strategies to improve the quality of vector stocks that we have presented here should be helpful in optimizing the transduction of biologically relevant cell types. Moreover, such strategies will significantly reduce the cost of vector production, making clinical trials of combinatorial RNAi strategies feasible.

Acknowledgments

RNAi research in the Berkhout laboratory is sponsored by ZonMw (VICI and Translational gene therapy grant) and NWO-Chemical Sciences (TOP grant). We thank Stef Heynen for the CA-p24 protocol.

References

1. Grimm, D., and Kay, M.A. (2007) Combinatorial RNAi: a winning strategy for the race against evolving targets? *Mol Ther* **15**, 878–88.

2. Liu, Y.P., and Berkhout, B. (2008) Combinatorial RNAi strategies against HIV-1 and other escape-prone viruses. *Int J Biosci Technol* **1**, 1–10.

3. Anderson, J., Li, M.J., Palmer, B. et al. (2007) Safety and efficacy of a lentiviral vector containing three anti-HIV genes-CCR5 ribozyme, tat-rev siRNA, and TAR decoy-in SCID-hu mouse-derived T cells. *Mol Ther* **14**, 1182–8.

4. Henry, S.D., van der Wegen, P., Metselaar, H.J., Tilanus, H.W., Scholte, B.J., and van der Laan, L.J. (2006) Simultaneous targeting of HCV replication and viral binding with a single lentiviral vector containing multiple RNA interference expression cassettes. *Mol Ther* **14**, 485–93.

5. Ter Brake, O., 't Hooft, K., Liu, Y.P., Centlivre, M., von Eije, K.J., and Berkhout, B. (2008) Lentiviral vector design for multiple shRNA expression and durable HIV-1 Inhibition. *Mol Ther* **16**, 557–64.

6. Song, J., Pang, S., Lu, Y., and Chiu, R. (2004) Poly(U) and polyadenylation termination signals are interchangeable for terminating the expression of shRNA from a pol II promoter. *Biochem Biophys Res Commun* **323**, 573–8.

7. Braun, S.E., Shi, X., Qiu, G., Wong, F.E., Joshi, P.J., and Prasad, V.R. (2007) Instability of retroviral vectors with HIV-1-specific RT aptamers due to cryptic splice sites in the U6 promoter. *AIDS Res Ther* **4**, 24.

8. Mitta, B., Rimann, M., and Fussenegger, M. (2005) Detailed design and comparative analysis of protocols for optimized production of high-performance HIV-1-derived lentiviral particles. *Metab Eng* **7**, 426–36.

9. Kumar, M., Keller, B., Makalou, N., and Sutton, R.E. (2001) Systematic determination of the packaging limit of lentiviral vectors. *Hum Gene Ther* **12**, 1893–905.

10. Ter Brake, O., and Berkhout, B. (2007) Lentiviral vectors that carry anti-HIV shRNAs: problems and solutions. *J Gene Med* **9**, 743–50.

11. Ter Brake, O., Konstantinova, P., Ceylan, M., and Berkhout, B. (2006) Silencing of HIV-1 with RNA interference: a multiple shRNA approach. *Mol Ther* **14**, 883–92.

12. Westerhout, E.M., and Berkhout, B. (2007) A systematic analysis of the effect of target RNA structure on RNA interference. *Nucleic Acids Res* **35**, 4322–30.

13. Seppen, J., Rijnberg, M., Cooreman, M.P., and Oude Elferink, R.P. (2002) Lentiviral vectors for efficient transduction of isolated primary quiescent hepatocytes. *J Hepatol* **36**, 459–65.

14. Brummelkamp, T.R., Bernards, R., and Agami, R. (2002) A system for stable expression of short interfering RNAs in mammalian cells. *Science* **296**, 550–3.

15. Yu, J.Y., DeRuiter, S.L., and Turner, D.L. (2002) RNA interference by expression of short-interfering RNAs and hairpin RNAs in mammalian cells. *Proc Natl Acad Sci USA* **99**, 6047–52.

16. Koper-Emde, D., Herrmann, L., Sandrock, B., and Benecke, B.J. (2004) RNA interference by small hairpin RNAs synthesised under control of the human 7SK RNA promoter. *Biol Chem* **385**, 791–4.

17. Denti, M.A., Rosa, A., Sthandier, O., De Angelis, F.G., and Bozzoni, I. (2004) A new vector, based on the PolII promoter of the U1 snRNA gene, for the expression of siRNAs in mammalian cells. *Mol Ther* **10**, 191–9.

18. Dull, T., Zufferey, R., Kelly, M. et al. (1998) A third-generation lentivirus vector with a conditional packaging system. *J Virol* **72**, 8463–71.

19. Kotsopoulou, E., Kim, V.N., Kingsman, A.J., Kingsman, S.M., and Mitrophanous, K.A. (2000) A Rev-independent human immunodeficiency virus type 1 (HIV-1)-based vector that exploits a codon-optimized HIV-1 gag-pol gene. *J Virol* **74**, 4839–52.

20. Andersson, M.G., Haasnoot, P.C.J, Xu, N., Berenjian, S., Berkhout, B., and Akusjarvi, G. (2005) Suppression of RNA interference by adenovirus virus-associated RNA. *J Virol* **79**, 9556–65.

21. de Vries, W, Haasnoot, J., van der Velden, J. et al. (2008) Increased virus replication in mammalian cells by blocking intracellular innate defense responses. *Gene Ther* **15**, 545–52.

22. Cambi, A., Gijzen, K., de Vries, J.M. et al. (2003) The C-type lectin DC-SIGN (CD209) is an antigen-uptake receptor for *Candida albicans* on dendritic cells. *Eur J Immunol* **33**, 532–8.

23. Haasnoot, J., de Vries, W., Geutjes, E.J., Prins, M., de Haan, P., and Berkhout, B. (2007) The Ebola virus VP35 protein is a suppressor of RNA silencing. *PLoS Pathog* **3**, e86.

24. Popa, I., Harris, M.E., Donello, J.E., and Hope, T.J. (2002) CRM1-dependent function of a cis-acting RNA export element. *Mol Cell Biol* **22**, 2057–67.

25. Liu Y.P., Haasnoot, J., and Berkhout, B. (2007) Design of extended short hairpin RNAs for HIV-1 inhibition. *Nucleic Acids Res* **35**, 5683–93.

Chapter 15

Lentivirus-Expressed siRNA Vectors Against Alzheimer Disease

Kevin A. Peng and Eliezer Masliah

Abstract

Amyloid precursor protein (APP) has been implicated in the pathogenesis of Alzheimer disease, and the accumulation of APP products ultimately leads to the familiar histopathological and clinical manifestations associated with this most common form of dementia. A protein that has been shown to promote APP accumulation is β-secretase (β-site APP cleaving enzyme 1, or BACE1), which is increased in the cerebrospinal fluid in those affected with Alzheimer disease. Through in vivo studies using APP transgenic mice, we demonstrated that decreasing the expression of BACE1 via lentiviral vector delivery of BACE1 siRNA has the potential for significantly reducing the cleavage of APP, accumulation of these products, and consequent neurodegeneration. As such, lentiviral-expressed siRNA against BACE1 is a therapeutic possibility in the treatment of Alzheimer disease. We detail the use of lentivirus-expressed siRNA as a method to ameliorate Alzheimer disease neuropathology in APP transgenic mice.

Key words: Alzheimer disease, Amyloid precursor protein, β-secretase, BACE1, Lentivirus, siRNA

1. Introduction

Alzheimer disease (AD), first described in 1901 by Alois Alzheimer, is a heretofore incurable and degenerative disease, affecting 26.6 million people in 2006. As such, it is the most common cause of dementia, and its prevalence is steadily increasing; some have estimated the prevalence to quadruple by 2050 (1).

The amyloid hypothesis for AD holds that accumulation of amyloid beta (Aβ) deposits, derived from amyloid precursor protein (APP), leads to the neurodegeneration and clinical presentation of AD (2). This was substantiated by experiments with transgenic mice expressing high levels of human APP, which showed that these APP transgenic mice developed Alzheimer-like

Maurizio Federico (ed.), *Lentivirus Gene Engineering Protocols*, Methods in Molecular Biology, vol. 614,
DOI 10.1007/978-1-60761-533-0_15, © Humana Press, a part of Springer Science+Business Media, LLC 2010

neuropathology (3, 4). It was demonstrated that amyloid beta is derived from amyloid precursor protein APP through the activity of β-secretase (β-site APP cleaving enzyme 1, or BACE1) and γ-secretase, which cleave APP at the β and γ locations, respectively (5). Based on this pathophysiology, it is intuitive to conceive that inhibiting the production and accumulation of APP products through inhibition of the secretases will delay or even halt the progression of AD.

Of these two secretases, β-secretase is more specific for the cleavage of APP (6). While β-secretase knockout mice are viable (7), mice lacking γ-secretase function did not survive, at best, past the peripartum period (8). These considerations alone made targeting β-secretase more promising than γ-secretase, an inclination that was supported by noting that specifically, BACE1 expression and generation of APP fragments is increased in AD (9–11). It was thus realized that BACE1 was indeed an attractive therapeutic target for AD.

Short, interfering RNAs (siRNAs) are segments of RNA between 21 and 23 nucleotides in length that silence transcripts containing complementary sequences. More specifically, siRNAs themselves are derived from short hairpin RNAs (shRNAs), which contain the siRNA sequence and are cleaved by cellular enzymes to produce the functional siRNA. When integrated into the host genome, as occurring through the lentiviral vector mediated transfection, these shRNAs can stably and persistently suppress expression of target RNAs (12).

We studied the effects of downregulation of BACE1 using siRNA against BACE1 in in vivo models of AD, specifically the previously mentioned APP transgenic mice. Lentiviral vectors were the delivery system of choice as they efficiently enter and express in nondividing cell types, such as neurons (13). More specifically, seven different sequences potentially targeting the human BACE1 open reading frame were inserted in different lentiviral vectors. One sequence (siBACE1-6) was found to efficiently reduce BACE1 expression in both human and mouse cells, and was used for all in vitro and in vivo experiments. As control, a lentiviral vector expressing siRNA against the glucose transporter 4 (siGLUT4) was used. Upon the intracerebral injection of siRNAs-expressing lentiviral vectors, we observed that lenti-siBACE1 delivered to APP transgenic mice resulted in decreased BACE1 expression compared with controls, and that, most importantly, lenti-siBACE1 reduced Alzheimer-like neuropathology (Fig. 1), (14). Here, we detail the use of lentivirus-expressed siRNA as a method to ameliorate of Alzheimer disease neuropathology in APP transgenic mice.

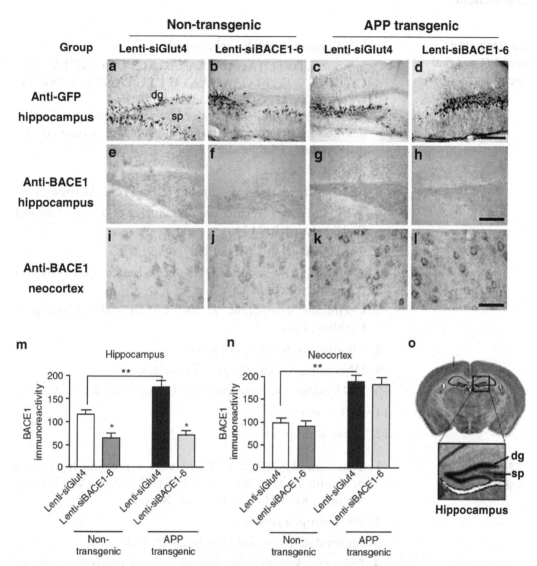

Fig. 1. Characterization of the effects of lenti-siBACE1-6 expression in the brains of APP transgenic mice. (a–d) Anti-eGFP immunoreactivity (lentiviral vectors express eGFP) shows consistent expression of lenti-siRNA constructs in nontransgenic and transgenic mice. (e–f) Nontransgenic mice show decreased BACE1 immunoreactivity with lenti-siBACE1-6 as compared with the control, lenti-siGLUT4. (g–h) APP transgenic mice treated with lenti-siBACE1-6 showed a dramatic reduction in BACE1 expression as compared to APP transgenic mice treated with the control, lenti-siGLUT4. (i–j) BACE1 expression in the neocortex of nontransgenic mice, where no injections were placed. (k–l) BACE1 expression in the neocortex of transgenic mice, where no injections were placed. (m) BACE1 immunoreactivity in the hippocampus ($P < 0.05$, compared with lenti-siGLUT4-treated controls, by one-way ANOVA with post-hoc Tukey–Kramer). (n) BACE1 immunoreactivity in the neocortex ($P < 0.05$, compared with nontransgenic controls, by one-way ANOVA with post-hoc Tukey–Kramer). (o) Anatomic representation of physical locations of the site of injection as well as of the neocortex. Reproduced from Singer *et al.*, 2005, with permission from *Nature Neuroscience*

2. Materials

2.1. Generation of shRNA and Viral Vectors

1. DNA oligonucleotide, gel purified

   5′-phos-CTGTCTAGACAAAAA**GACTGTGGCTACAA-CATTC**<u>TCTCTTGAA</u>**GAATGTTGTAGCCACAGTC**GGGGATCTGTGGTCTCATACA-3′

   (Bold face denotes sense and antisense sequences; underline denotes loop sequence; terminal plain face sequences are discussed in of Subheading 3.1.1).

2. 20×SSC.

3. pLentilox-derived lentivirus cloning vector (ATCC, Manassas, VA).

4. *Xba*I and *Nhe*I restriction endonucleases (New England Biolabs, Ipswich, MA).

5. 1% agarose gel in TBE buffer.

6. T4 DNA ligase and 10×DNA ligase buffer (New England Biolabs, Ipswich, MA).

7. Chemically competent *E. coli* DH5α cells (Invitrogen, Carlsbad, CA).

8. LB plates containing ampicillin.

9. DNA miniprep kit (Qiagen, Valencia, CA).

10. DNA midiprep or maxiprep kit (Qiagen, Valencia, CA).

11. 94°C water bath.

12. Sterile toothpicks.

2.2. Intracerebral Injection

1. Transgenic C57BL/6×DBA/2 mice expressing high levels of double-mutant human APP751 (London (V717I) and Swedish (K670M/N671L)) (see Note 1).

2. 3% Hydrogen peroxide.

3. Antiviral detergent solution (e.g., 1% Virkon S).

4. Protective clothing (coat, disposable gloves, protective eyewear).

5. Animal facility.

6. Gas or injection anesthetics for small animals

   a. 30% O_2 + 70% N_2O mixture containing 1.5–2% isoflurane or halothane, or

   b. Sodium pentobarbital.

7. Small animal stereotaxic frame (David Kopf Instruments, Tujunga, CA).

8. Surgical microscope.

9. Bone drill.

10. Hamilton syringe, 5 µl, with 26-gauge steel cannulas.

11. Suture kit and surgical instruments.

2.3. Tissue Preparation and Immunohistochemical Transgene Detection

1. 4% w/v paraformaldehyde fixative, pH 7.0 to 7.4 at 4°C.

2. 0.9% w/v NaCl in deionized water, or phosphate-buffered saline (PBS), at room temperature.

3. 20–30% w/v sucrose in deionized water.

4. 2 mm O.D. perfusion cannula connected to an adjustable flow pump.

5. Cryostat (Leica Microsystems, Bannockburn, IL) or freezing microtome (Leica Microsystems, Bannockburn, IL).

6. 12-well plates.

7. Quenching solution: 10% v/v methanol + 3% v/v hydrogen peroxide in PBS.

8. Incubation buffer: 2% v/v normal serum + 0.25% v/v Triton x-100.

9. Primary antibodies

   a. Aβ (1:2000; Signet Laboratories, Dedham, MA).

   b. MAP2 (see Note 2; 1:50; Chemicon, Billerica, MA).

   c. BACE1 (1:500; ProSci, Poway, CA).

10. Secondary antibodies

    a. FITC-conjugated (for detection of Aβ and APP CTFs).

    b. Biotinylated goat anti-rabbit (for detection of BACE1).

11. Blocking serum: normal goat serum.

12. Vectastain ABC kit (Elite PK-6101, Vector Laboratories, Burlingame, CA).

13. 3′3′-diaminobenzidine and 3% hydrogen peroxide color reaction kit (DAB; Sigma-Aldrich, St. Louis, MO).

14. Ethanol, diluted serially (100%, 95%, 70%, 50%, 20%).

15. Histoclear (Fisher Scientific, Pittsburgh, PA) and 1:1 mixture of Histoclear and ethanol.

16. Entellan (Sigma-Aldrich, St. Louis, MO).

17. Glass slides and cover slips (Fisher Scientific, Pittsburgh, PA).

3. Methods

3.1. Design and Generation of siRNA and Vectors

3.1.1. Design of shRNA Oligonucleotides

To express siRNAs in transduced cells, first insert shRNA sequences under the control of an RNA polymerase III (Pol III) promoter. The 83-mer shRNA sequences, specified above, contain the following.

1. A 19-nucleotide sense strand (the siRNA against BACE1) and the corresponding 19-nucleotide antisense strand, separated by a nine-nucleotide loop.

2. A five-adenine sequence serving as a template for the Pol III promoter termination signal.

3. 20 nucleotides complementary to the 3′ end of the Pol III H1 promoter.

4. A 5′ end containing an *Xba*I restriction site.

For a discussion of siRNA and shRNA design for arbitrary target sequences, see Note 3.

3.1.2. Generation of shRNA

1. PCR is used to amplify fragments containing the entire H1 promoter plus the 83-mer shRNA construct detailed above. Cycles used are the following.

   a. Initial cycle of 94°C for 3 min

   b. 30 cycles of 94°C for 30 s

   c. 55°C for 40 s

   d. 72°C for 50 s

2. Cut the resulting product with *Xba*I and ligated to an *Nhe*I-digested lentivirus vector at a molar ratio of 3:1.

3. Transform the ligated mixture into chemically competent DH5α cells according to the manufacturer's instructions. Plate onto LB plates containing the appropriate antibiotic (e.g., ampicillin for pLentilox vectors).

4. Using a sterile toothpick, pick colonies and isolate miniprep DNA using a miniprep kit. Digest approximately 1 μg of vector DNA with *Xba*I and *Nhe*I, and electrophorese a sample on an agarose gel (15).

5. Sequence two to three plasmids to identify successfully inserted vectors.

6. Prepare midipreps or maxipreps of the DNA containing the correct sequence, as well as the lentiviral packaging vectors, using appropriate kits.

3.1.3. Production of Lentiviruses Expressing shRNA

Lentiviral vector production and concentration protocols are detailed comprehensively in previous chapters of this volume. Briefly, human embryonic kidney 293 cells containing the SV40 T antigen (HEK293T cells) are transfected via the calcium phosphate method with plasmids encoding lenti-siBACE1 as well as plasmids encoding for the production of lentiviral packaging proteins. Lentiviruses are harvested by filtering the supernatant at 48 and 72 h through a 0.22-μm-pore cellulose acetate filter, followed by centrifugation and ultracentrifugation. Quantification of vector concentration should be performed (see Note 4). For this experiment, lentiviral aliquots at 7.5×10^9 TU/ml were used (see Note 5).

3.2. Animals and Intracerebral Injection

For this experiment, we use transgenic mice expressing high levels of double-mutant human APP751 (London (V717I) and Swedish (K670M/N671L) under control of murine Thy1 regulatory sequences. These mice have been shown previously to develop neuronal amyloid plaques consistent with AD neuropathology (3, 4). Experimental mice are injected unilaterally with lentiviruses containing lenti-siBACE1 into the hippocampus, and control mice are injected unilaterally with lentiviruses containing lenti-siGLUT4. Throughout, we use Franklin and Paxinos atlas coordinates (16, 17).

1. Prepare lentiviral vector aliquots (see Note 6).

2. Anesthetize the animal using a gas or injection anesthetics.

3. Place anesthetized animal on a stereotaxic apparatus and make an incision in the skin over the skull. Dissect away connective tissue until the bregma of the skull is visible. Drill a hole in the skull bone at the appropriate anteroposterior and mediolateral coordinates relative to the bregma.

4. Load 2 μl of virus (i.e., 1.5×10^7 TU) into a 5-μl Hamilton syringe. Lower the tip of the needle until it touches the dura and then read the dorsoventral coordinates. Lower the needle to the appropriate dorsoventral coordinates.

5. Using the Hamilton syringe, inject the virus at 0.25 μl/min into the hippocampus. Wait 5 min before retracting needle (see Note 7).

6. Suture or clip skin incision.

7. Animals may survive for at least four weeks following injection.

3.3. Tissue Preparation and Immunohistochemical Transgene Detection

Incubations are performed in well plates placed on a shaker. Alternatively, 20-ml glass or plastic vials may be used.

1. Anesthetize mice with chloral hydrate and flush-perfuse transcardially with 0.9% saline.

2. Remove brains and postfix in phosphate-buffered 4% PFA (pH 7.4) at 4°C for 48 h (see Note 8).

3. Cut 40-μm-thick sections with a vibratome.

4. Immunolabel sections with antibodies against Aβ or APP c-terminal fragments (CTFs), followed by incubation with fluorescein isothiocyanate–(FITC)-conjugated secondary antibodies; rinse. Alternatively, if measuring BACE1 expression, immunolabel sections with antibodies against BACE1, followed by incubation with biotinylated anti-rabbit secondary antibodies; rinse (see Note 9).

5. Incubate with ABC according to the manufacturer's instructions; rinse.

6. Prepare DAB and hydrogen peroxide solution according to the manufacturer's instructions; develop color and stop reaction by rinsing in $1 \times$ PBS.

7. Mount sections onto glass slides.

8. Dehydrate in increasing concentrations of ethanol (20%, 50%, 70%, 95%, and 100%), followed by a 1:1 mixture of Histoclear and ethanol, followed by Histoclear.

9. Add one drop of Entellan rapid embedding agent on each glass slide and coverslip.

10. Image with a laser scanning confocal microscope (LSCM) (see Note 10).

4. Notes

1. We have described the production of transgenic mice extensively (18). (C57BL/6 and DBA/2 F1 mice may be obtained from Charles River, Hollister, CA, or The Jackson Laboratory, Bar Harbor, ME.)

2. Microtubule-associated protein 2 (MAP2) is a marker of dendritic integrity, and can be used to quantify the extent of neurodegeneration.

3. The process of selecting siRNA sequences, given a target mRNA sequence, is discussed in an excellent reference available through the Tuschl laboratory (http://www.rockefeller.edu/labheads/tuschl/sirna.html). BLAST (www.ncbi.nlm.nih.gov/BLAST) may be used to confirm that the target sequences of interest do not share extensive homology (> 15 bp) with sequences not of interest in the experimental species. It is recommended that several target sequences be chosen, since it is extremely difficult to predict the efficacy of any particular siRNA sequence. Instead, the efficacy of each siRNA sequence may be validated by Western blot analysis of cell cultures transfected with vectors containing each siRNA sequence.

4. Vector concentrations may be analyzed using immunocapture p24-*gag* ELISA or by flow cytometry of transduced HEK293T cells, as previously described (13, 19).

5. Vector concentrations should exceed 1×10^8 TU/ml to ensure small volume of delivery and maximal tissue concentration.

6. Keep the lentiviral aliquots on ice. These should be mixed throughout surgery to maintain consistent concentration.

7. If the needle is removed immediately after injection completes, the vector may track into the vacant path of the needle rather than diffuse into target tissue. Therefore, it is advised to wait 5 min following injection prior to withdrawing the needle.

8. If desired, tissues may be fixed in paraffin. Sections would then be cut 5–10 μm thick, and sections would be placed on slides and serially hydrated (Histoclear, 1:1 Histoclear and ethanol mixture, ethanol at 100%, 95%, 70%, 50%, 20%, PBS) prior to staining with standard immunohistochemistry protocols.

9. For other applications, primary and secondary antibodies may be chosen as per user preferences.

10. To estimate amyloid load, percent area of neuropil occupied by Aβ-immunoreactive plaques may be used. To evaluate integrity of neuronal structure, percent area of neuropil occupied by MAP2-immunoreactive dendrites and synaptophysin-immunoreactive terminals may be used.

References

1. Brookmeyer, R., Johnson, E., Ziegler-Gram, K., and Arrighi, H. M. (2007) Forecasting the global burden of Alzheimer's disease *Alzheimer's and Dementia* **3**, 186–91.

2. Hardy, J., and Allsop, D. (1991) Amyloid deposition as the central event in the aetiology of Alzheimer's disease. *Amyloid deposition as the central event in the aetiology of Alzheimer's disease* **12**, 383–8.

3. Games, D., Adams, D., Alessandrini, R., Barbour, R., Borthelette, P., Blackwell, C., Carr, T., Clemens, J., Donaldson, T., Gillespie, F., Guido, T., Hagopian, S., Johnson-Wood, K., Khan, K., Lee, M., Leibowitz, P., Lieberburg, I., Little, S., Masliah, E., McConlogue, L., Montoya-Zavala, M., Mucke, L., Paganini, L., Penniman, E., Power, M., Schenk, D., Seubert, P., Snyder, B., Soriano, F., Tan, H., Vitale, J., Wadsworth, S., Wolozin, B., and Zhao, J (1995) Alzheimer-type neuropathology in transgenic mice over-expressing V717F beta-amyloid precursor protein. *Nature* **373**:523–7.

4. Masliah, E., Sisk, A., Mallory, M., Mucke, L., Schenk, D., and Games, D. (1996) Comparison of neurodegenerative pathology in transgenic mice overexpressing V717F beta-amyloid precursor protein and Alzheimer's disease. *J Neurosci* **16**, 5795–811.

5. Cai, H., Wang, Y., McCarthy, D., Wen, H., Borchelt, D. R., Price, D. L., and Wong, P. C (2001) BACE1 is the major (beta)-secretase for generation of A(beta) peptides by neurons. *Nat Neurosci* **4**, 233–4.

6. Selkoe, D. J., Yamazaki, T., Citron, M., Podlisny, M. B., Koo, E. H., Teplow, D. B., and Haass, C (1996) The Role of APP Processing and Trafficking Pathways in the Formation of Amyloid {Beta}-Protein. *Annals of the New York Academy of Sciences* **777**, 57–64.

7. Roberds, S. L., Anderson, J., Basi, G., Bienkowski, M. J., Branstetter, D. G., Chen, K. S., Freedman, S., Frigon, N. L., Games, D., Hu, K., Johnson-Wood, K., Kappenman, K. E., Kawabe, T. T., Kola, I., Kuehn, R., Lee, M., Liu, W., Motter, R., Nichols, N. F., Power, M., Robertson, D. W., Schenk, D., Schoor, M., Shopp, G. M., Shuck, M. E., Sinha, S., Svensson, K. A., Tatsuno, G., Tintrup, H., Wijsman, J., Wright, S., and McConlogue, L (2001) BACE knockout mice are healthy despite lacking the primary {beta}-secretase activity in brain: implications for Alzheimer's disease therapeutics. *Hum Mol Genet* **10**: 1317–24.

8. Shen, J., Bronson, R., Chen, D., Xia, W., Selkoe, D., and Tonegawa, S. (1997) Skeletal and CNS Defects in Presenilin-1-Deficient Mice. *Cell* **89**, 629–39.

9. Holsinger, R. M. D., McLean, C. A., Beyreuther, K., Masters, C. L., and Evin, G. (2002) Increased expression of the amyloid precursor beta-secretase in Alzheimer's disease *Annals of Neurology* **51**, 783–6.

10. Fukumoto, H., Cheung, B. S., Hyman, B. T., and Irizarry, M. C. (2002) {beta}-Secretase Protein and Activity Are Increased in the Neocortex in Alzheimer Disease. *Arch Neurol* **59**, 1381–9.

11. Zetterberg, H., Andreasson, U., Hansson, O., Wu, G., Sankaranarayanan, S., Andersson, M. E., Buchhave, P., Londos, E., Umek, R. M., Minthon, L., Simon, A. J., and Blennow, K. (2008) Elevated Cerebrospinal Fluid BACE1 Activity in Incipient Alzheimer Disease. *Arch Neurol* **65**, 1102–7.

12. Stewart, S. A., Dykxhoorn, D. M., Palliser, D., Mizuno, H., Yu, E. Y., An, D. S., Sabatini, D. M., Chen, I. S. Y., Hahn, W. C., Sharp, P. A., Weinberg, R. A., and Novina, C. D. (2003) Lentivirus-delivered stable gene silencing by RNAi in primary cells. *RNA* **9**, 493–501.

13. Naldini, L., Blomer, U., Gallay, P., Ory, D., Mulligan, R., Gage, F. H., Verma, I. M., and Trono, D. (1996) In vivo gene delivery and stable transduction of nondividing cells by a lentiviral vector. *Science* **272**, 263–7.

14. Singer, O., Marr, R. A., Rockenstein, E., Crews, L., Coufal, N. G., Gage, F. H., Verma, I. M., and Masliah, E. (2005) Targeting BACE1 with siRNAs ameliorates Alzheimer disease neuropathology in a transgenic model. *Nat Neurosci* **8**, 1343–9.

15. Crittenden, J. R., Heidersbach, A., and McManus, M. T. (2007) Lentiviral Strategies for RNAi Knockdown of Neuronal Genes. In: Current Protocols in Neuroscience: John Wiley and Sons.

16. Rosenblad, C., and Lundberg, C. (2003) Cells of the Nervous System. In: Federico M, ed. Lentivirus Gene Engineering Protocols. Totowa, NJ: Humana Press; 299–307.

17. Franklin, K. B. J., and Paxinos, G. (2007) The Mouse Brain in Stereotaxic Coordinates. 3 ed: Academic Press.

18. Rockenstein, E., Mallory, M., Mante, M., Sisk, A., and Masliah, E. (2001) Early formation of mature amyloid-beta protein deposits in a mutant APP transgenic model depends on levels of Abeta *Journal of Neuroscience Research* **66**, 573–82.

19. Marr, R. A., Rockenstein, E., Mukherjee, A., Kindy, M. S., Hersh, L. B., Gage, F. H., Verma, I. M., and Masliah, E. (2003) Neprilysin Gene Transfer Reduces Human Amyloid Pathology in Transgenic Mice *J Neurosci* **23**, 1992–6.

INDEX

Printed in the United States
By Bookmasters

Printed in the United States
By Bookmasters